# Science of Electricity

# Volume 6

# Coal Power Technologies Explained Simply

by Mark Fennell

© 2012

This book is part of the
Energy Technologies Explained Simply™ Series

# Other Books in the Energy Technology Series

# About the Book

Overview

    This book discusses everything which decision-makers and citizens need to know about Coal Power. The major topic areas include: Coal Mining Operations, Mining Safety, Operation of Coal Power Plants, Environmental Effects, and Clean Coal Technologies.

Importance of this Book

    In our society today many aspects of coal power are vigorously debated. However, too often people with an agenda present lies about coal power as absolute truth. Too often the policy-makers make bad decisions based on emotions or on myths rather than on evidence. Too often the citizens are bombarded with conflicting information, leaving them confused. That is why this book is so important. This book provides the unbiased, accurate truths of all aspects related to coal power.

    I had no agenda when researching and writing this book. I never had any preconceived theories to prove or disprove.

    Certainly I had questions (the same questions you do). Therefore I spent a significant amount of time to find the answers. I embarked on an expedition, a quest, to learn the science, to understand the technologies, and to separate truth from fiction, for all aspects of coal power.

    In this book you now have the summation of my journey. In this book you will find the accurate, unbiased answers to all your questions regarding coal power technologies.

Accuracy of this Book on Coal Power

    I never relied solely on the conclusions of other researchers. Instead, I performed many other tasks to ensure that all conclusions were accurate.

    Note that my background is in Chemistry, and I have used all of my technical knowledge to analyze each aspect of coal power.

    I examined primary data whenever possible. I have read the fine print on how research was obtained. I have also checked the accuracy of the conclusions written by other researchers, most commonly by finding at least three distinct sources for each fact.

In addition, I performed my own calculations numerous times to prove (or disprove) conclusions and final data found in various reports.

It is only after such rigorous investigations that I created data tables and wrote summaries. Because of the labor I have performed in fact checking, I fully stand by every statement and every data item in this book.

Information Not Found Anywhere Else

In addition to the accurate information, this book also has information which is not found in any other book, journal, or website. Some of this information includes:

1) Mining Safety: The latest mining safety techniques, from experts in West Virginia and Kentucky. This includes techniques to reduce Black Lung, which are not yet known to the highest levels of the Mine Safety Health Administration.

2) Temperature Data: You have heard about "global warming" – now see some actual temperature data. Find temperature data here that you will not find in other books.

3) Mercury Emissions: The science of mercury emissions from coal power plants has only recently been studied. The data is also scattered. This is the first book which collects all the latest science behind mercury emissions and mercury capture.

4) Global Temperature: An entire chapter is devoted to the issue of coal power and environmental temperature. This is a complex topic, with many factors. In this book you will find a clear explanation of each factor, combined with a chemical engineer's analysis of each.

5) This book also explains the information very clearly. There are many scientific concepts, including aspects of chemistry, physics, and biology. All of these concepts are explained very clearly, in such way that any reader will be able to understand, regardless of background.

# Chapter Contents

This book discusses all aspects of coal power. Major topic areas include: Coal Mining Operations, Mining Safety, Operation of Coal Power Plants, Environmental Effects, and Clean Coal Technologies.

Chapter one discusses formation and types of coal; energy from coal; and the basic operation of coal power plants.

Chapter two discusses coal mining methods, including details of underground mining, surface mining, and mining reclamation.

Chapter three discusses coal mining safety. This chapter was written from a personal perspective – that is, if I were working in the mine or if a family member was working in the mine, I would want the mining operations as safe as possible. Detailed areas of mining safety include: Black Lung, Collapsing Mines, Flooding, and Ventilation.

Chapter four discusses the environmental effects of coal power. The most common complaints of coal power are related to the environmental effects. Here you will learn all the negative effects of coal power on the environment. We will also discuss the most effective methods to minimize each of those environmental effects.

In chapter four we will also discuss each Clean Coal Technology. These clean coal technologies include: scrubbers, low NOx burners, Selective Catalytic Reduction (SCR), Electrostatic Precipitator, Circulating Fluidized Bed Combustion (CFBC), and Gasification.

Chapter five is devoted to $CO_2$ and the environment. This is a complex topic, involving a variety of scientific concepts. All concepts are discussed and explained. In addition, this chapter provides actual temperature data.

Chapter six, the final chapter, is devoted to mercury emissions from coal power. This is a relatively new topic among researchers, with scattered data and emerging scientific discoveries. We will examine the types of mercury in coal, the types of mercury being emitted, and the most effective ways of capturing mercury.

We will also discuss the hazards of various mercury compounds and the relative mercury hazards from coal power plants.

The study of mercury emissions from coal power plants is relatively new. Data is scattered among various research organizations. Technology for mercury capture is currently being developed and tested. This book is the first book which gathers all the latest science of this topic in a single reference.

## Fact-Based Conclusions

Note that while I have never had an agenda, I do have conclusions. These conclusions are based on the facts I have carefully researched and analyzed. I will not be shy about stating these fact-based conclusions.

For the record, I am now a general supporter of coal power. I had no opinions either way at the beginning of the project, but the facts have led me to where I stand today.

In the beginning of my research, I thought I might find that coal power was bad for the environment, that clean coal technologies were inadequate, and that coal power contributed to global warming. Instead, after studying the data, I have learned that coal power can be clean. I have learned that coal power can coexist with the environment. This is not what I believe – this is what the data shows.

It is important that you know that all my conclusions throughout this book which appear to support specific areas of coal power have come from detailed and diligent research.

It was not prior opinions nor a prior agenda to prove. Nor was it simply reading the words of the coal power industry and accepting those words at face value. All statements, data, and conclusions which appear to be in support of coal power in any way have come from detailed research and analysis, using a variety of sources, many of which are opposed to coal power or have no affiliation with coal power.

All of this information will be discussed and presented scientifically throughout the book.

# About the Energy Technology Series

<u>Purpose of this Series</u>

The books in the *Energy Technologies* series are designed to educate citizens, students, and legislators on all aspects of energy technologies. The first books in the series focus on electrical power.

The books discuss many energy technologies, including: generators, turbines, power plants, power lines, and grids. The technologies for each type of power source (hydro, wind, solar, coal, nuclear, and natural gas) are discussed in detail. The books also discuss efficiency, safety, reliability, and health concerns for each energy technology.

The ultimate goal of the series is to enable the people to make informed decisions on practical energy questions. The secondary goal is to serve as introductory guides for students embarking on careers with energy technologies.

Taken altogether, the books in the series answer any question you are likely to have, such as:
- How can we increase the efficiency of solar cells?
- How do I select the size my solar array?
- What do I need to know when installing a wind turbine?
- How effective are the clean coal technologies?
- How can we prevent grid failures?
- Do power lines cause cancer?
- and many other energy technology questions...

<u>Science of Electricity in Perspective</u>

The subject of electrical power is of great importance to our communities, but is rarely taught. Public debate is frequent and passionate, but with too little understanding of the actual science. At best, an informed citizen knows only a few pieces. At worst, as it is for a great number of citizens, electricity is magic and myths are believed as scientific truth. It does not have to be that way. Any citizen, regardless of background, can know the technologies behind all aspects of electricity.

The books in this series solve that problem. These books educate the general public in all aspects of electrical power. Any person, regardless of background, can easily find the answer to his energy question in one of these books.

## Specific Goals

There are numerous technologies described in these books. Yet for each technology I sought out the answers to the following questions:

1. How does the technology work?
2. What are the advantages and disadvantages?
3. What is the efficiency? How can the efficiency be improved?
4. What is the environmental impact? How can it be improved?
5. What are the safety hazards, and how can they be reduced?
6. What are the most important practical tips?
7. What facts comprise the most important data?

## Technical Discussions Explained Simply

The books in the series must necessarily be technical to some degree. Electricity is a practical technology, and therefore we must understand the technical aspects if we want to make wise decisions. Yet the discussions in this book are always aimed at the citizen or policy maker.

The books in this series explain the principles of electricity as simply as possible, using ordinary English (no engineering jargon), and highlighting the most important points of each technology. Main concepts and facts are emphasized with the use of lists, tables, diagrams, and summaries.

I do not expect any reader to have a background in science, yet I offer enough facts and details so that the reader can have an accurate understanding of all related technologies. I provide enough technical details and enough data for the reader to make informed decisions.

## Conclusion

For all the reasons above, I offer this series of books. My goal is to inform you so that you can make realistic decisions. Remember there are no perfect solutions, there are only choices. I hope that this series of books will assist you in making those choices for your community.

Mark Fennell

# Table of Contents

# Table of Contents: Detailed

# 6.1
# Coal Power Basics

## Introduction

Coal is a major player in the production of electricity. This is for two reasons. First, coal was the first fuel to be used to generate electricity. Second, there is enough coal in America to be used for generations to come.

The United States has a large supply of coal reserves. Estimates place the amount of coal as enough supply to last us for 200-300 years. This is one of the primary advantages to using coal for our electricity.

The efficiency of coal power plants is 35% – 40%. This value is up from the extremely low efficiency of 5% a century ago. Estimates on future efficiency claim up to 55-70%, particularly with gasification (a recent technology) and with dual turbines.

Note that technology has improved over the years to make coal a much friendlier fuel than is commonly thought. If a coal power plant is built today with the most advanced clean coal technologies then 90%-95% of ash and other pollutants will be collected.

List of topics for this chapter
1. Formation and Types of Coal
2. Energy and Power of Coal
3. Basic Operation of the Coal Power Plant
4. Turbines in Coal Power Plants
5. Abbreviations to Know for Coal Power

# Formation and Types of Coal

## Introduction

Millions of years ago, the earth was covered with plants and swamps. After these plants died they were buried in the swamp. As years went by, the plants were further buried under land and additional dead plants. The continuing process of the burial created high pressures and high temperatures. Over millions of years, these high pressures and high temperatures converted the elements within the plants into the form known as coal.

Coal is a natural substance made primarily of Carbon and Hydrogen. Coal also usually contains significant quantities of Sulfur and Nitrogen. Other elements include Silicon, Aluminum, Iron, and Calcium. The specific composition of the coal depends on the region where the coal was created.

The classification of coal types is based on the heating value, which is defined as "the amount of energy per mass of coal". This heating value is directly related to the age of the coal, and secondarily related to the location of the coal.

## Types of Coal

There are four types of coal: Lignite, Sub-Bituminous, Bituminous, and Anthracite. The classification of coal is based on several factors. These factors are: Progression of Time; Carbon Content; Moisture Content; and Heating Value. All of these factors are inter-related.

These factors can be understood as follows: As time progresses, one form of coal turns into the next form. This means that the moisture content decreases, the non-carbon elements disappear, and the coal becomes a higher percentage of only carbon. A higher percentage of carbon means that the coal will have more energy per mass of coal.

All these factors together (higher percentage of carbon; lower moisture content; and fewer non-carbon elements) result in the later forms of coal having a greater heating value than the previous forms.

Any of these types of coal, including Lignite, may be used for electrical power plants. However, the greater moisture content makes the lower forms of coal less economical to transport (due to the added weight of the water).

Therefore, if the lower stages of coal are to be used (lignite, sub-bituminous), these forms of coal should be used in close distances to where they are mined.

## 1. Lignite
- Lignite is the lowest form of coal.
- Lignite has the lowest Carbon content.
- Lignite has high moisture content: as high as 45%
- Lignite has a high ash content.
- Lignite has the lowest heating value: 9-17 million BTU per ton.
- Lignite is a brown colored coal.

## 2. Sub-Bituminous
- Sub-Bituminous has a better heating value than lignite, yet still has high moisture content.
- Sub-Bituminous heating value: 17-24 million BTU per ton.
- Sub-Bituminous has low Carbon content.
- Sub-Bituminous has high moisture content: 20-30%
- Sub-Bituminous has a high ash content.
- Sub-Bituminous is a dark brown or black coal.

## 3. Bituminous
- Bituminous coal is the most commonly used coal because it has a relatively high heating value and relatively low moisture content.
- Bituminous coal has a good heating value: 19-30 million BTU per ton.
- Bituminous has a relatively high Carbon content.
- Bituminous coal has low moisture content: 20% or less.
- Bituminous is a black coal or dark brown coal.

## 4. Anthracite
- Anthracite has the highest heating value of all the coals. Anthracite is also the cleanest of all coals.  However, anthracite is the least plentiful.
- Anthracite has the highest heating value: 22-28 million BTU per ton.
- Anthracite is nearly pure carbon: 90% carbon.
- Anthracite has the lowest moisture content: less than 15%.
- Anthracite has the lowest ash content.
- Anthracite is the hardest and blackest of all coals.

## Energy and Power of Coal

The energy of coal is measured in "Millions of BTU per Ton of Coal." This is also known as the "heat content" of the coal. The energy in coal can range from 9 million BTUs to 28 million BTUs per ton of coal. The average heat content of the most commonly used type of coal is approximately 25 million BTUs per ton of coal. This is equivalent to approximately 7,500 kw-hrs from each ton of coal.

The power from coal is measured in units of BTUs per hour. The amount of power we can get from coal depends on several factors. The most important factors are 1) the type of coal and 2) how fast we burn the coal.

Other factors include 3) which region the coal came from, 4) the preparation process, and 5) the type of furnace used.

We can get an approximate idea of the amount of power we can get from coal by assuming that we use the most common type of coal (Bituminous), and that we burn one ton of coal in one hour. This gives an approximate value of 25 million BTUs/hr, or 7,500 kilowatts.

## Coal Power Plant Basic Operation

Basic process (Figure 6.1)

The basic process of using coal for generating electricity is relatively simple: Coal is burned, which then creates heat. This heat is used to boil water, turning the water into steam. The steam pushes the blades of the turbine, which then operates the electrical generator.

The steps in the production of power from burning coal can be broadly categorized into 5 sets of processes:

A. Preparation
B. Burning
C. Turbines and Generating Electricity
D. Reusing Water
E. Removal of By-Products

A. Preparation process:
    1. Coal is washed.

      This eliminates one form of sulfur, as well as eliminating dirt and some other minerals which would become ash.

    2. Coal is pulverized into small pieces.

      Turning the coal into small pieces gives more surface area to the coal overall, and thus burns more efficiently.

B. Burning process:
    3. The coal is put into the furnace.
    4. Air is let in at a constant stream which allows a steady burn.
    5. The coal is burned, usually at 1,500 to 2,500 degrees Fahrenheit.

C. Steam turbine and electrical generator process:
    6. The heat from the burning of coal is sent past water, boiling that water into steam.
    7. The steam is sent to push the turbine blades.
    8. The turbine operates the generator which creates the electricity.

D. Reusing water:
    9. Water is cooled and sent back to be reheated again.

E. By-products removal process:
    10. By-products are created in the furnace:
        a. Gas phase by-products: $CO_2$, $H_2O$, $SO_x$, $NO_x$
        b. Ash (solid phase products)
        c. Small particles of ash float through the pipes

    11. Most of the ash settles to the bottom.

      The ash which settles to the bottom is collected easily, and it never reaches the atmosphere.

    12. The gases and smaller ash particles continue through the pipes.

      The pipe is technically the "flue" and these gases are the "flue gases."

13. The smaller ash particles are collected.

These smaller ash particles are collected by an electrostatic precipitator. This collection of ash is easily removed, and thus is not put into the atmosphere.

14. Nitrogen by-products are collected.

There are many methods to reduce the amount of these pollutants. The most common is Selective Catalytic Reduction (SCR).

15. Sulfur by-products are collected.

There are many methods to reduce the amount of these pollutants. The most common method is Flue Gas Desulfurization (FGD). The Sulfur removal step is often referred to as the "scrubber."

16. The only by-products reaching the outside air are carbon dioxide and water.

Note that you will often see white vapor coming from the stacks, yet this vapor is not smoke, nor is this vapor harmful to the environment. This vapor is essentially water vapor. This vapor may also include a small amount of limestone (introduced at the sulfur removal step).

17. Remaining by-products are collected and sold.

The ash, sulfur by-products, and nitrogen by-products are often sold to various industries. These by-products can also be buried without any danger.

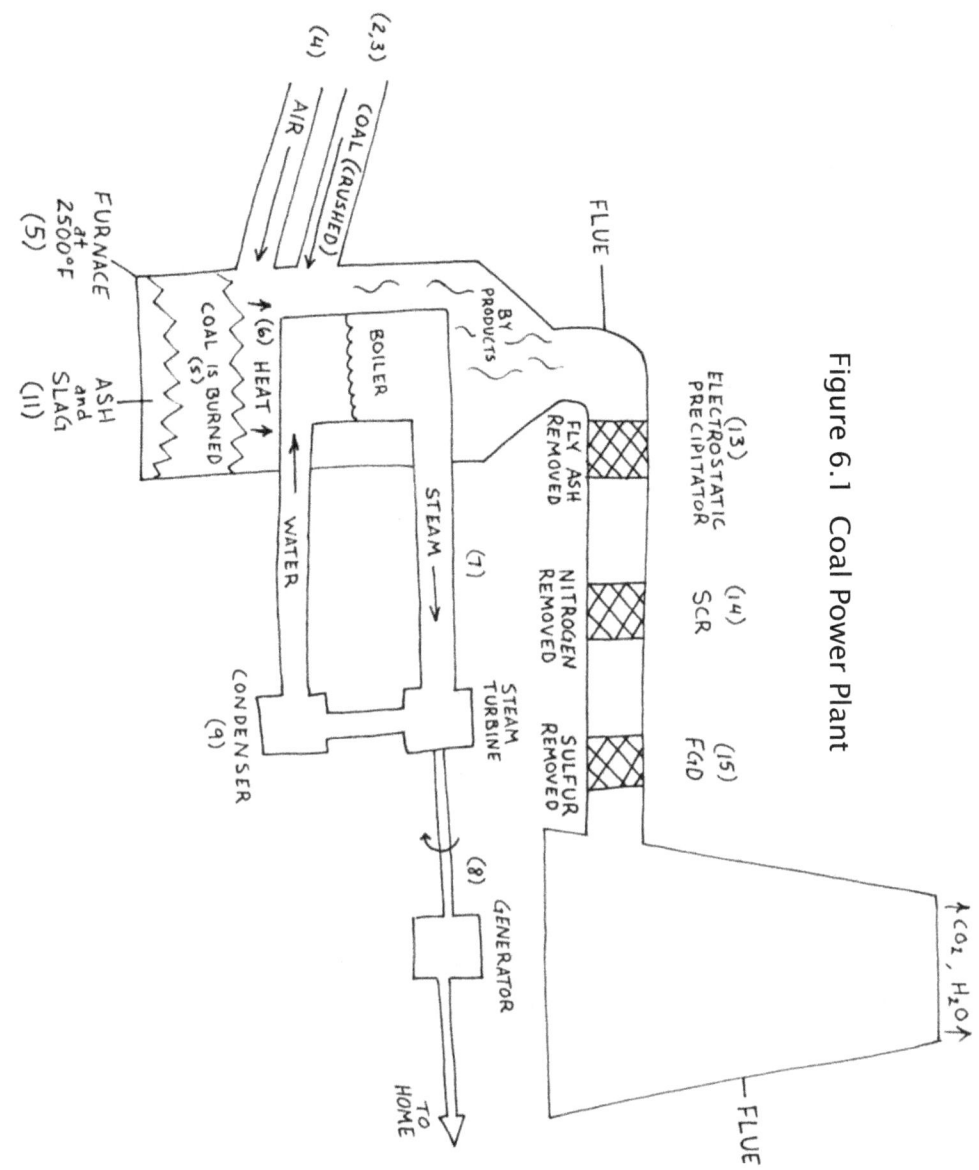

Figure 6.1 Coal Power Plant

## Turbines in Coal Power Plants

Two types of turbines

When we burn coal, we create two things which can be used to operate turbines: 1) heat (energy released from combustion), and 2) molecules in the gas phase (by-products of combustion). Therefore, coal power can use two different types of turbines: steam turbine or gas turbines.

In a steam turbine, the heat from the burning process is applied to water, which then boils into steam. The steam pushes on the turbine blades and causes the turbine to rotate. This is the most traditional type of turbine.

In a gas turbine, the by-product molecules push the turbine blades. The by-product molecules from coal combustion are usually carbon dioxide ($CO_2$), water ($H_2O$), sulfur dioxide ($SO_2$), and nitrogen dioxide ($NO_2$). These gas phase molecules push the turbine blades, causing the turbine to rotate.

These two turbines can be combined. The gas molecules are used first, operating the gas turbine. The energy leftover is then used as heat to boil water into steam, and thus we use the steam turbine. Using two turbines allows us to generate more electricity for the same amount of coal burned.

## Abbreviations to Know

The coal power industry uses many abbreviations. The following guide will assist you when you read information from other sources.

CCP: Coal Combustion Products

A Coal Combustion Product is any by-product from the burning of coal which can be sold for use in other industries. Ash is the most significant Coal Combustion Product, but there are also forms of Sulfur and Nitrogen that can be sold.

CCT: Clean Coal Technology

Clean Coal Technology is any technology which makes burning of coal more environmentally friendly.

FGD: Flue Gas Desulfurization

Flue Gas Desulfurization is the primary means of removing $SO_2$. The basic operation is to spray a limestone-water slurry into the flue. This slurry captures the $SO_2$, which can then be taken away. Flue Gas Desulfurization is commonly referred to as a scrubber.

IGCC turbine: Integrated Gasification Combined-Cycle system

The IGCC is a dual turbine, which means that two turbines are used in sequence. The first turbine is a gas turbine, where the gas molecules created by the combustion push the turbine blades. The second turbine is a traditional steam turbine, where the heat from the combustion boils water into steam. The IGCC system allows us to get more power from the coal, which results in greater efficiency.

SCR: Selective Catalytic Reduction

Selective Catalytic Reduction (SCR) is the primary technology to reduce the amount of $NO_2$ molecules. The basic process involves adding Ammonia, which converts the $NO_2$ molecules into water and pure Nitrogen.

## Chapter Summary

1. Coal is created from decaying plants that died millions of years ago.

2. There are four types of coal: Lignite, Sub-Bituminous, Bituminous, and Anthracite.

3. The types of coal are successive stages of the coal formation process.

4. The earlier forms of coal have more water; lower carbon content; and lower heating value. The later forms of coal have less water; higher carbon content; and higher heating value.

5. There is enough coal in America to be used for generations to come.

6. Coal is a much cleaner fuel than is commonly thought.

7. Coal power can use two turbines, the steam turbine and/or the gas turbine. These turbines can be used together as well as individually.

8. Coal and the steam turbine: burning coal creates heat which is used to boil water, turning the water into steam. The steam pushes the turbine, which in turn operates the generator.

9. Coal and the gas turbine: burning coal creates gas products. These gas products push the turbine, which operates the generator.

10. Common abbreviations associated with coal power are:
    a. CCP: Coal Combustion Products
    b. CCT: Clean Coal Technology
    c. FGD: Flue Gas Desulfurization
    d. IGCC turbine: Integrated Gasification Combined-Cycle turbine
    e. SCR: Selective Catalytic Reduction

11. The efficiency of coal power plants is 35%–40%. Estimates on future efficiency claim up to 55%–70%.

12. Energy of coal is measured in "Millions of BTU per Ton of Coal." This is also known as the "heat content" of the coal.

13. The average heat content of coal is approximately 25 million BTUs per ton of coal, which is approximately equivalent to 7,500 kw-hrs.

# 6.2
# Modern Coal Mining Methods

## Introduction
Mines are usually categorized as either surface mines (commonly known as strip mines) or underground mines. Mines are also sometimes referred to by the type of entry. Mines are further classified by the mining technique used. In this chapter we will look at the most important concepts related to coal mining, including methods, hazards, and best practices.

List of topics for this chapter
1. Surface Mining
2. Reclamation of Surface Mines
3. Underground Mining
4. Types of Entry

## Surface Mining

Introduction
Surface mining generally requires careful removal of the topsoil, digging the coal from the earth, and returning the topsoil to the mine.

There are three types of surface mines: Area Surface Mines, Contour Mines, and Open Pit Mines. However, for our purposes the distinctions are not important.

There are several reasons to choose surface mining rather than underground mining. First, surface mining is safer than in underground mining. There are no chances of cave-ins, no flooding, and there is plenty of oxygen. Methane gas, coal dust, and silica are spread into the air, not confined, and therefore are not as hazardous to the workers.

Second, surface mining is more complete than in underground mining. With underground mines, there must be some method of supporting the roof, which means some coal must be left in place. Yet with surface mining, all the coal can be taken from the area.

## General Process

A large excavating machine removes the top layer of soil, called "overburden," and sets it aside. This top layer must be removed in order to reach the coal below. The coal is then dug out by large shovels and put into large trucks to be hauled away.

Surface mining didn't get its start until a machine was created which could remove the overburden effectively. This machine, called the "dragline," is an enormous machine. Some people say it is the largest machine on land today. The bucket alone can carry 250 tons. The trucks which haul the coal are also very large. Each truck is several times larger than any common construction vehicle, and is capable of holding several hundred tons of coal.

All surface mines are reclaimed when done. This means that the hole is filled in, the topsoil replaced, and the trees replanted. After the land has been reclaimed, it is almost impossible to tell that a mine was ever there.

## Blasting in Surface Mines

Another method to remove the overburden is by using explosives. Explosives remove overburden quickly, yet will cause much damage to the environment. When a company uses explosives for removing overburden, they are essentially blowing the tops off of the mountains.

These explosions are blasting hillsides on an enormous scale, which causes problems to all areas nearby the mine. Debris from the explosion fills the rivers, toxic chemicals enter the drinking water, and ecosystems are destroyed forever. Therefore, compromises must be made between the mining companies and the local communities.

Environmental impact can be minimized by using smaller explosives and by directing the blasts away from the streams. However, the very nature of explosions will always transform the mountains significantly.

One alternative to blasting in surface mines is transitioning to wind power. In this method wind turbines are placed on the mountain tops (which is a very effective location for wind power). The mountains remain untouched and there is no environmental impact. At the same time the power company is still able to make profit – they are just using a different type of power. Additionally, the wind turbines provide power to local communities in a very efficient manner.

# Reclamation of Surface Mines:
# The Surface Mining Reclamation Act of 1977

## Introduction

Strip mining is a very effective way to get coal. It is much safer for men to get coal, for there are no ventilation issues, and no concern for cave-ins. However, surface mining will leave a large hole, often several miles long and hundreds of feet deep. Because of this fact, a law was passed in 1977: the Surface Mining Control and Reclamation Act. This law requires that the hole be filled in, the top soil replaced, and trees replanted. The law also specifically states that the slope of the reclaimed mine must be no greater than 10% difference from the original slope of the land.

## Reclaiming the Surface Mine

In the surface mining process, much of the rock and dirt from the top must be taken out. In the past, this overburden was just thrown to the side. Today, this rock and dirt is put back into the pit of the strip mine. Companies take care to keep the original topsoil separate, so that same topsoil can be put on top at the very end. In addition, mining companies replant trees and other vegetation before moving on.

Since the passing of the law, many surface mines have been reclaimed. These areas have been successfully turned into habitats for wildlife, into local parks, and into sites for homes and businesses. Many of these areas are now so full of grassland, trees, and wildlife that it is not obvious that there was ever a surface mine at that location.

## Reclamation of Blasting Mountains

Note that areas which have been blasted away by large explosions cannot be reclaimed in the same way that strip mines can. After blasting the topsoil and trees are at the bottom of the mountains and in the local rivers, and cannot be returned to original locations.

In addition, the regulations and enforcement operations of blasting operations are not held to the same standards as enforcement for strip mines. This is a situation which requires further investigation by local communities and their representatives.

# Underground Mining

## Introduction

Underground coal mining requires: breaking the coal off the walls into chunks, transporting these large chunks of coal out of the mine, and supporting the walls of the mines in order to prevent collapse.

There are three primary methods for underground mining: conventional, continuous, and long wall. We will discuss each method.

## Conventional Mining

Conventional mining uses explosives to break the coal off of the wall. The miners drill a hole, put in an explosive, and the explosive breaks the coal from the wall. The coal chunks can be taken from the mine in a variety of means, typically by a shuttle car.

In order to prevent the roof of the mine from caving in, a method is used called "Room and Pillar Mining." Large pillars of coal are left standing throughout the mine, and each pillar supports the roof of the mine. This has the effect of creating many "rooms"; hence the term Room and Pillar Mining (Figure 6.2).

Sometimes coal is also taken from these pillars. This can only be done when coal has been taken from the entire level of the mine. Taking coal from pillars must be done carefully because the roof of the mine will eventually collapse. The miners determine beforehand how many pillars can come down safely before the roof collapses.

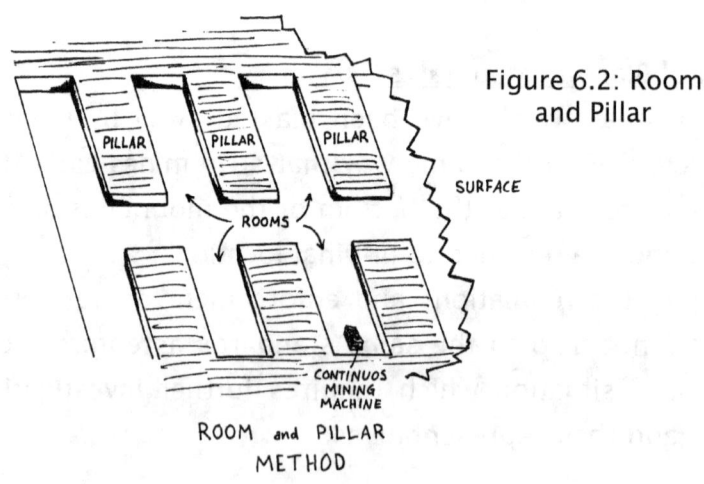

Figure 6.2: Room and Pillar

## Continuous Mining (Figure 6.3)

Continuous Mining uses a machine called a "Continuous Miner." This machine cuts at the wall, removing chunks of coal. This machine replaces the drilling and exploding done in conventional mining. It also does the work much faster than humans could ever do.

This machine is operated with remote control by an operator standing just a few feet away. The Continuous Miner has a rotating steel drum with carbide teeth. The carbide teeth cut into the wall, breaking the coal off the wall into chunks. Most of these Continuous Miner machines also automatically transfer the coal chunks onto a conveyor belt or to shuttle cars. As with Conventional Mining, the method of Continuous Mining requires the use of "Room and Pillar" methods so that the roof does not collapse.

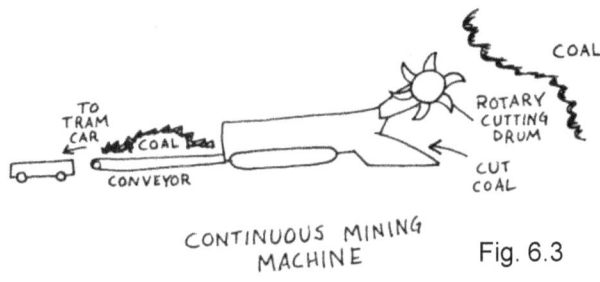

CONTINUOUS MINING MACHINE          Fig. 6.3

## Long Wall Mining (Figure 6.4)

A Long Wall Machine is a very long machine, easily several hundred feet long, and sometimes a full mile in length. This machine has numerous cutting devices along the length of the machine. The long wall machine cuts the coal, then marches forward, cuts more coal, and then marches forward again. It is a methodical cutting and marching forward, much like a hundred men cutting coal and moving forward as a single unit.

The long wall machine also moves the coal automatically, putting the coal either onto conveyor belts or into a shuttle car.

Fig. 6.4

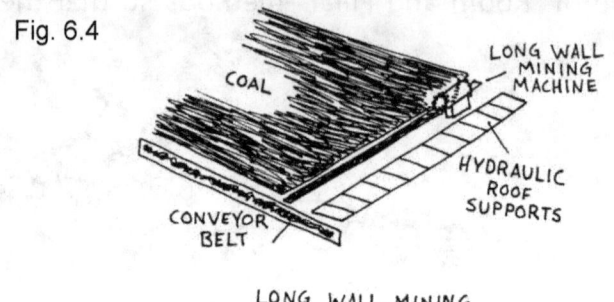

LONG WALL MINING

The method of roof support for Long Wall Mining must be different than for other mining methods. Because of the nature of Long Wall Mining, there are no pillars to support the roof and prevent collapse. Therefore, the roof is supported by a hydraulic shield which is part of the Long Wall Machine.

As the wall is cut away the roof is supported by a hydraulic roof support, known as the "shield." As the Long Wall Machine moves forward, this shield progresses behind the machine, supporting the roof over the mined area.

Behind the shield, the roof is allowed to collapse in a controlled manner. As the Long Wall Machine moves forward, the end of the shield is also moved forward, and the roof above that excavated area collapses. Thus, in Long Wall Mining, the roof over a mined area is allowed to collapse in a very slow and very controlled manner.

# Types of Entry

Underground mines are also sometimes classified by the types of entry. There are three types of entry to underground mines: Drift, Slope, and Shaft (Figure 6.5). The different types of entrances are based on the depth required to get to the coal. Of the entry types, drift mines have coal nearest to the surface, shaft mines have coal that is the deepest, and slope mines have coal which lies somewhere in between.

The angles to the entrance is also different for each type of entry. Drift mines have essentially horizontal entrances. Slope mines have sloped entrances. Shaft mines have elevators, which take the miners straight down deep into the mine.

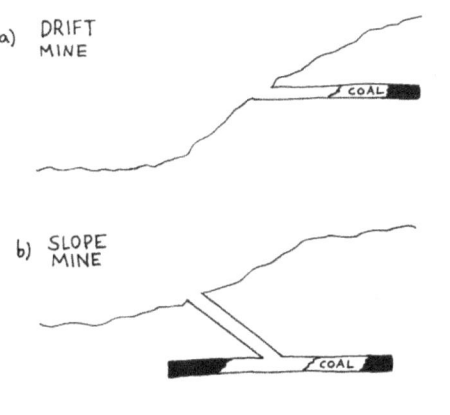

Drift mines have essentially horizontal entrances.

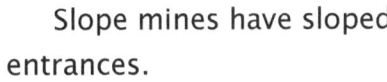

Slope mines have sloped entrances.

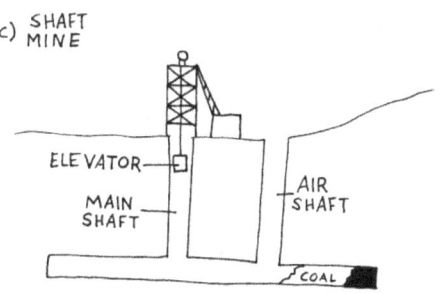

Shaft mines have elevators, which take the miners straight down, deep into the mine.

Fig. 6.5 Types of Mines by Entry

27

# Chapter Summary

1. Mines can be classified as Surface or Underground.

2. Surface Mines are commonly called Strip Mines. Surface mining is safer than underground mining. A greater percentage of coal can be removed through surface mining than from underground.

3. According to The Surface Mining Control and Reclamation Act of 1977, mining companies are required to fill in the hole created by mining and to replant grass and trees on the topsoil.

4. Underground mining requires these steps:
   a. Breaking the coal off the walls into chunks
   b. Carrying these large chunks of coal out of the mine
   c. Supporting the walls of the mines in order to prevent collapse

5. Underground mines can be classified either by the method of breaking the coal off the walls, or by the type of entry.

6. The types of underground mines, classified by the method of breaking the coal off the walls, are: Conventional; Continuous; and Long Wall
   a. Conventional: using explosives or picks.
   b. Continuous: using a machine called a "Continuous Miner" which has large carbide teeth.
   c. Long Wall: using a long machine, often a mile in length, that picks at the wall and moves forward

7. In order to support the roof in an underground mine, the following methods are used: room and pillar; shields; or roof bolts.

8. The types of entry for underground mines are: Drift; Slope; and Shaft
   a. Drift: horizontal entrance
   b. Slope: angled entrance
   c. Shaft: vertical entrance; these are deep mines requiring elevators to reach the coal

# 6.3
# Coal Mining Safety

## Introduction

If we are to use coal power, then we must provide a safe work environment for the miners. There are five main concerns regarding health issues of coal mining: Collapsing Mines, Flooding, Need for Oxygen, Methane Explosions, and Black Lung. In this chapter we will discuss the health issues of mining, and methods to reduce those risks.

List of topics for this chapter:
1. Collapsing Mines
2. Water and Flooding
3. Ventilation Problems (Oxygen and Methane)
4. Ventilation Solutions
5. Black Lung
6. Dust Control
7. Face Masks and Respirators

## Collapsing Mines

The first task for any mining company is to ensure that the mine will not collapse as the workers dig through the mine. There are three ways to prevent the mines from caving: 1) Pillars of Coal, 2) Shields, and 3) Roof Bolts. Pillars and shields were discussed earlier. Roof bolts are exactly as the term implies. A roof bolting machine puts long bolts into several layers of strata in the roof of the mine. The weaker layers of rock are held in place by a stronger layer further up, with the bolts holding the layers together.

## Water

Water naturally flows underground in many areas of the country. Whenever a miner digs underground it is likely that he will encounter water at some point in the mine. Breaking a hole in the wrong spot has been known to cause a large amount of flooding, which can trap the miners.

Water can also make the soil weak, making cave-ins more likely. In order to prevent these events, engineers must study the local water tables carefully, and the water must be pumped out before mining begins. Furthermore, in most mining operations the water must be pumped out during the entire mining process.

## Ventilation

There are two main problems requiring ventilation: 1) providing oxygen for breathing, and 2) reducing the methane in order to prevent explosions.

In any underground mine, oxygen will be limited. After securing the mine to prevent collapse, the mining company must build ventilation shafts so that the miners can breathe comfortably.

Methane is very combustible. In fact, we use methane as fuel for many purposes. Obviously, methane is not desirable inside a mine because of the potential for explosions.

Methane gas is produced along with the creation of coal. Therefore, where there is coal, there is methane. When the mine is open to air from the outside, Oxygen comes in. Methane plus Oxygen, in a warm mine, can create an explosion.

Adding to this problem is coal dust. Remember that when we burn coal in power plants we deliberately pulverize the coal into smaller pieces. Smaller pieces of coal burn more efficiently. However, we have this same situation, naturally, in the coal mines. After the methane starts burning inside the mine, this heat starts the coal dust burning. The burning coal dust adds to the problem, and creates a great explosion. Therefore, it is very important when mining underground to remove both methane gas and coal dust so that there are no explosions.

Note that any form of ventilation has two components: blowing in, and drawing out. Regarding oxygen, the main goal is to bring fresh air into the area. Regarding methane explosions, the goal is to remove the methane and coal dust from the area. To help provide adequate ventilation to the workers, techniques include the following: ventilation shafts, large fans, oxygen tanks, respirators, reasonable shifts, methane detectors, and oxygen detectors.

# Black Lung

Coal miners have long had to deal with the health problem commonly referred to as "black lung." The cause of black lung is silica. In most places where you find coal, you will also find silica. The silica particles are constantly in the air of the mine, and the particles are very fine. Miners breathe in this silica easily, and over a period of years the silica damages the lung.

The silica particles are troublesome to remove because they are very fine, and will pass through most filters. The particles are too fine to get rid of completely. Modern engineering methods have reduced the fine particles which cause black lung by 60% since 1970. However, black lung is still a serious problem. We must continue to find engineering techniques which will reduce the amount of silica particles breathed by the miners.

# Dust Control

There are many specific methods for dust control, yet these methods can be organized into a few basic approaches:
1. Collect and Remove the dust
2. Cover the dust
3. Wet the dust
4. Move the worker further from the dust

Specific Solutions for Dust Control include the following:
1. Install large scale ventilation
2. Spray coal chunks with water or foam
3. Spray the walls and ceiling with limestone slurry
4. Use machinery to keep men away from the dust
5. Install spray mechanisms in mining equipment
6. Design dust collection mechanisms into each mining machine
7. Maintain devices related to dust reduction
8. Cover specific areas
9. Chop the coal into the largest chunks possible
10. Avoid dropping the coal
11. Allow miners to work proper hours for shifts
12. Use face masks and respirators

For each of these methods, we will note some tips and cautions.

## 1. Use ventilation systems to remove silica and coal dust from the area

These exceptionally large ventilation units not only provide air to the miners, but also remove many of the dust particles. Air velocity is a factor. The higher the velocity, the more oxygen and the more dust moved through a system. Regulations require a minimum air velocity of 60 feet/minute. Recommended air velocity should be between 60 and 150 ft/minute.

Caution: If the ventilation system is not designed properly, then the ventilation system will cause *more* harm than without it. Remember that most of these areas are small rooms (recall the section on room and pillar mining.) If air is blown on the dust in the wrong way, then the dust will hit the walls of the mine and blow back into the face of the miner. Studies have shown that several improperly designed air velocity systems have *increased* the amount of dust a worker was exposed to, not decreased the amount as was intended.

## 2. Spray the coal with water or foam

a. Spray the coal chunks with water: Water holds the dust particles to the main piece of coal, resulting in less dust floating around in the mine. There are two important techniques to make spraying water most effective: 1) wet the coal as it is being broken off the wall, and 2) wet the coal uniformly.

b. Spray the chunks of coal with foam: Foam is more effective at controlling dust than water. Spraying the coal with foam tends to be 30%–50% more effective at reducing dust than spraying water at the same locations. The only disadvantage to spraying with foam is the higher cost.

## 3. Spray the walls and ceiling of mine with limestone slurry

Spray the ceiling, walls, and floor of the mine with a limestone slurry: The limestone sticks to the ceiling and floor, and therefore prevents most of the dust from floating in the air.

## 4. Use more machinery in order to keep men away from the dust

Use more machines to mine the coal, such as the continuous miner and the longwall mining machine. When machines are used to chop the coal instead of a man, then there will be fewer men inside the mine. Thus, with fewer men inside the mine, fewer men will be exposed to the silica particles.

Furthermore, the men that are in the mine will be further from the dust than if the men were at the wall themselves. A continuous miner allows the worker to stay 12 feet away, which reduces his exposure to dust by 50%. A longwall mining machine allows the operator to stay 15–20 feet away, which reduces his exposure to dust by 68%.

## 5. Install spray mechanisms in mining equipment

Spraying as a method of dust control is most effective when done as the coal is being chopped. Install small spray mechanisms into the coal mining machines. The spray can be water, foam, limestone slurry, or some other liquid. When the spray mechanism is part of the mining machine the spraying occurs efficiently. Furthermore, additional sprays (pointing in other directions) can wet the general area of the mine. Spraying the area can bring the particles down and out of the air immediately after the coal has been chopped from the wall.

## 6. Design dust collection mechanisms into each mining machine

Machines cut coal from the wall at a fast pace. In the process, particles of all sizes come off, from large chucks of coal, to fine particles invisible to the eye. However, devices can be built into the workings of the machine to reduce the amount of fine particles produced which reach the miners. The most effective of these devices are the vacuum and trap device. This dust collection mechanism is in essence a vacuum cleaner, sucking up the dust as soon as it is formed, and storing it inside the machine. The trap will need to be emptied periodically.

Caution: We must not forget that this method of dust collection is in essence a large vacuum cleaner. Remember that if the vacuum-trap system is not emptied on regular schedules (which can be quite often), then it will be useless. We must also dispose of the dust properly. It does not do any good to capture the dust, only to dispose of it in a place where it can be blown around again.

Furthermore, the mining machine chops coal of the wall at a fast rate. This results in a high rate of dust being created. The vacuum must have the proper suction, and the trap must have a large enough volume, to effectively remove this quantity of dust.

## 7. Maintain every device related to dust reduction

Clean and repair every device related to dust reduction, as often as needed. Maintenance should be done not just on mining machines, but on also devices such as roof bolting systems, ventilation systems, conveyer belts, spray mechanisms, and traps.

One of the most overlooked methods of dust control is to adequately maintain the dust control equipment. NIOSH stated this problem most clearly: "Screens and filters clog often, sometimes more than once per shift. Gaskets disappear and access doors leak. Often, filters are not seated properly, and dusty air leaks around them. Filters develop holes from mishandling and from abrasion by larger size particulate. Ductwork leading to the collector fills with coarse particulate, cutting off the airflow. Fans located on the inlet side of the collector suffer rapid erosion of their blades."

Therefore it is very important to properly maintain all dust control equipment. Every component related to dust control must be properly inspected, cleaned, and replaced, preferably on a predictable schedule.

## 8. Cover specific areas with curtains, tents, and tubes

Use a method of covering over specific areas to reduce the amount of dust reaching the workers. Just as a tent covers a house during fumigation, various coverings can be built over the machines and working areas which can keep the particles from reaching the miners beyond. In addition, these coverings can be made of the highest quality filters which can prevent the finest particles from coming through.

There are three basic coverings or dividers: a) the curtain, b) the tent, and c) the tube. The curtain is the simplest, and can be put up quickly to separate a worker from the coal dust. Large tents can be used to cover machines, such as the continuous miner, while chopping the coal. Tubes can be used to surround conveyer belts; any dust falling from the coal while exiting will be contained in the tube.

Before the worker enters any covered space, the dust can be actively filtered, sprayed down, or simply be allowed to settle.

## 9. Cut the coal from the wall in the largest chunks possible

Cutting off larger chunks will produce less dust. There are two factors for cutting coal in large pieces: a) deep cuts, and b) sharp blades. Deep cuts are best achieved by using the longest blades possible. Sharp blades are best achieved by sharpening the blades or replacing the blades whenever necessary.

## 10. Avoid dropping the coal

Every time a piece of coal is dropped, some of that coal will break. Inevitably, some of the pieces will be very tiny, become airborne, and therefore can be inhaled by the workers. Therefore we should avoid dropping coal as much as possible.

Coal will break from a drop into smaller pieces with either a higher fall or a harder landing surface. Therefore, tips include:
  a. Avoid dropping coal from great heights.
  b. Cushion the fall of coal whenever possible.
  c. Where major drops must occur and cushioning is not possible, use adequate covering, wetting, and/or ventilation as needed.

## 11. Allow miners to work proper hours for shifts

Allow miners to work a reasonable number hours, allowing sufficient time away from the dust particles between their shifts. After we have done all that we can to mitigate the fine dust particles, we should look at the remaining health risks, and compute a proper shift schedule.

## 12. Use face masks and respirators

After all other dust control methods have been put into place, use face masks or respirators as final protection for each worker. (The details of face masks and respirators will be discussed in the next section.)

# Face Masks and Respirators

## Introduction

After all other dust control methods have been put into place, use face masks or respirators as final protection for each worker. Ideally, when a miner wears a face mask, he will breathe only air, not coal dust or silica. However, there are factors to consider when making face masks or respirators as effective as they were designed to be. In addition, there are a few key legal requirements.

## Factors for Effective Use of Respirators

1. Respirators must be fitted to each person.
2. Use the highest quality filters available
3. Filters must be changed on schedule
4. The filter changing schedule is specific to each mine

(Each of these will be discussed briefly below)

## 1. Respirators must be fitted to each person.

The edges of the face mask or respirator must fit as closely to the skin as possible. An improper fit will allow dust to enter the sides of the mask. It is not possible to fit a mask perfectly 100%, but you can get close. Note that a worker might have to try several respirators, including respirators from different companies, before finding the one that fits his face best.

## 2. Use the highest quality filters available

Remember that the primary cause of black lung is silica, and remember that silica particles are usually very small. Therefore, the filters on the face masks and respirators should filter the smallest particles possible.

## 3. Filters must be changed on schedule

The face mask or respirator is useless if the filter is full. Therefore, every filter should be replaced on schedule. This schedule is specific to the type of filter and specific to the conditions in the mine.

## 4. <u>The filter changing schedule is specific to each mine</u>

Each schedule is specific to the circumstances of the mine and the particular respirator. Factors for determining the schedule include:

    a. concentration of silica in the mine

    b. rate of dust created (due to rate of mining)

    c. air velocity from ventilation systems

    d. capacity of the filter (# mg of dust that the filter will hold)

## <u>Legal Requirements and Respirators</u>

In addition to the tips above, there are some important legal requirements to be aware of regarding respirators:

1. Respirators must never be used instead of the engineered dust control methods listed above.

2. When silica concentration is at a certain level, respirators must be used. (The specific level is determined by MSHA.)

3. Respirators must be properly maintained and inspected.

4. Respirators must be fitted for each worker.

# <u>Chapter Summary</u>

1. There are four main concerns regarding health issues of mining: Collapsing Mines; Water; Ventilation; and Black Lung

2. In order to prevent the mines from collapsing, use Pillars of Coal, Shields, or Roof Bolts.

3. In order to prevent water from flooding the mine, pumping water is still the best method. In addition, engineers and miners must know the flow of underground water in the region.

4. Ventilation is needed to provide oxygen (for breathing), and to reduce the amount of methane (in order to prevent explosions).

5. Techniques for proper ventilation include:

    a. ventilation shafts

    b. large fans

    c. oxygen tanks

    d. respirators

    e. working in reasonable shifts

    f. using methane and oxygen detectors.

6. Black lung is caused by silica particles. In most places where you find coal, you will also find silica.

7. Preventing a worker from breathing the silica is difficult because the particles are very small, and they are constantly floating around in the mine.

8. The basic approaches to dust control are:

    a. removing the dust

    b. covering the dust

    c. wetting the dust

    d. moving the worker further from the dust

9. Important factors to make respirators effective include:

    a. Respirators must be fitted to each person

    b. Use the highest quality filters available

    c. Filters must be changed on schedule

    d. The schedule for changing filters must be determined for each location

10. Legal requirements of using face masks and respirators include:

    a. Respirators must never be used instead of the engineered dust control methods.

    b. When silica concentration is at a certain level (as determined by regulations), respirators must be used.

    c. Respirators must be properly maintained and inspected.

    d. Respirators must be fitted to each person.

# 6.4

# Environmental Effects of Coal Power Plants, and Clean Coal Technology

## Introduction

Coal power has become much cleaner due to the Clean Coal Technologies developed over the past 30 years. In this chapter we will look at the primary pollutants from coal power and discuss how these pollutants can be minimized. We will also discuss the details of Clean Coal Technologies.

List of topics for this chapter
1. Primary Pollutants and Methods to Minimize
2. Clean Coal Technologies
3. Gasification Details
4. Coal Plants Must be Modernized

## Primary Pollutants and Methods to Minimize Each

Introduction

The process of burning is a combination of any fuel with oxygen. The elements of choice for use as fuel are carbon and hydrogen. This means that the primary by-products of most fuels are carbon dioxide and water. However, other elements that are with the coal will also "burn", meaning that these elements will also react with oxygen and therefore create oxides of those other elements.

The primary concerns over coal power are the environmental effects of the by-products, such as ash, $SO_2$, $NO_x$, and $CO_2$. Therefore, in the following sections, we will examine the environmental effects of each by-product, and take a brief look at how each by-product can be collected, removed, and disposed. Details of each technique will be discussed in the section on Clean Coal Technologies.

## Ash (Solid By-Products)

Ash is any product from burning which is created in a solid form. The specific molecules in ash are very much like rocks and minerals found throughout the earth. The main elements which become ash are common metals such as: Silicon, Aluminum, Iron, and Calcium. When these elements burn they produce common oxides: $SiO_2$, $Al_2O_3$, $Fe_2O_3$, and $CaO$.

There are three main types of ash, depending on how and where they are collected. These types of ash are: bottom ash, slag, and fly ash. Bottom ash settles to the bottom of the furnace. Bottom ash can be cleared out of the furnace simply, and then taken away. Slag is ash that has been fused together by high temperatures into a material that is essentially coarse sand or glass. Like bottom ash, slag can be cleared out easily and taken away. Fly ash is a collection of smaller particles that floats in the pipes along with the gas phase by-products. This ash can be separated and collected easily by an electrostatic precipitator. The fly ash is then disposed of easily.

Ash can be separated and removed by several methods: 1) washing the coal, 2) removing bottom ash and slag (by hand or by machine), 3) collecting fly ash by an electrostatic precipitator, and 4) gasification. Details of these technologies will be explained in our discussion of Clean Coal Technology.

Disposing of ash is not a problem, because ash is used in other industries. There are many industrial uses for ash, most commonly as filler in concrete and as a base in paving material. There are also many companies which exist to use ash to its fullest potential. Therefore we do not need to be concerned with putting this ash in a landfill or burying it – every bit of this ash can be used.

Almost 100% of ash can be collected and not pollute the atmosphere. Most of the ash falls to the bottom of the furnace and is easily collected. All of the remaining ash ("fly ash") floats with the gasses (by-products from combustion). Almost all of this fly ash (99.7% of the fly ash) is collected by using electrostatic precipitators. The remainder of the fly ash is collected by scrubbers. Disposing the ash is not a problem because all the ash can be used, particularly in concrete and roads.

## Sulfur By-Products (SO$_x$)

Sulfur is found with coal in two forms. One form is in a separate material, pyrite, which is found with coal but not chemically attached. This form of sulfur can simply be washed off. Sulfur is also bound chemically to the coal. This sulfur must be burned along with the coal in order to remove. Burning sulfur produces various sulfur oxides. However, scrubbers can remove these molecules before they reach the atmosphere.

The primary by-product of burning sulfur is SO$_2$. However, there are multiple varieties of sulfur oxides. Therefore the total of all sulfur oxide by-products are abbreviated to SO$_x$.

Each of these SO$_x$ molecules exists in the gas phase, which means they naturally float into the air.

The major problem with SO$_x$ is acid rain. The SO$_x$ by-products from the burning of coal can combine with water (H$_2$O) in the air, and turn into sulfuric acid (H$_2$SO$_4$). This acid is then in our air. The acid can attack metal structures, and this acid can come down with the rain.

Note that we said *can* become acid; this process does not always occur. The amount of acid rain produced varies with the type of coal used, the amount of preparation of the coal before burning, humidity levels, wind patterns, and the general climate of the area.

Sulfur can be removed by several methods. These methods include: Flue Gas Desulfurization (FGD), gasification, and circulating fluidized bed combustion. Flue Gas Desulfurization removes 90%–97% of the SO$_x$.

## Nitrogen By-Products (NO$_x$)

Nitrogen is chemically bound to the coal, therefore this nitrogen must be burned with the coal and removed from the mix of by-products later. After coal is burned, a series of by-products known as NO$_x$ are created, all of which are in the gas phase.

Similar to sulfur, the main problem with NO$_x$ by-products is the potential for acid rain. Some of the NO$_x$ by-products will combine with water (H$_2$O) in the air, and turn into nitric acid (HNO$_3$). This acid is then in our air, corroding structures, and coming down with the rain.

Nitrogen can be removed by low temperature NOx burners, Selective Catalytic Reduction (SCR), and gasification. Details of each of these technologies will be explained in our discussion of Clean Coal Technology.

## Mercury

Mercury is far less of a problem than many people realize. There is no definitive study at this time which shows mercury to be a serious problem from coal power. Furthermore, the overall percentage of mercury in U.S. coal is only .0000075% (weight percent.) This means that 1 ton of coal, on average, contains only .0024 ounces of mercury.

Also, the amount of mercury released by coal power plants in the United States is only 1% of the mercury released into the environment worldwide.

For more information on Mercury emissions from coal power, see the related chapter in this unit.

## Carbon Dioxide

The effect which Carbon Dioxide has on the environment is one of the most passionate debates of energy technology. Therefore this topic is discussed in great detail later in this book. We can summarize the major scientific facts as follows:

1. Carbon Dioxide is not a pollutant.
2. All plants grow stronger with an increase in Carbon Dioxide.
3. Carbon Dioxide does have an effect on the temperature.
4. Carbon Dioxide is not the only factor which can change the temperature.
5. Temperature trends have always been cyclic.
6. The most effective way to capture Carbon Dioxide is to plant trees and other vegetation nearby.

For more details read the section on Carbon Dioxide later in this book.

# Clean Coal Technologies

## Introduction

Coal power has gotten a bad reputation due to its polluting operations many years ago. However, coal power plant technology has improved over the years. The process of burning coal today is much cleaner than it was in the past.

From 1970 to 2000, pollutants from coal power have been reduced significantly:

- $SO_2$ Reduction (1970-2000): 68%
- $NO_2$ Reduction (1970-2000): 45%
- Ash Reduction (1970-2000): 40%

In this chapter we will look at the various types of Clean Coal Technologies, and learn how modern technologies have eliminated most of the environmental problems.

## List of Clean Coal Technologies

1. Washing Coal
2. Scrubbers, including Flue Gas Desulfurization (FGD)
3. Low [temperature] NOx Burners (low temp., thus low NOx)
4. Selective Catalytic Reduction (SCR)
5. Electrostatic Precipitator
6. Circulating Fluidized Bed Combustion (CFBC)
7. Planting trees and other vegetation
8. Gasification

# Brief Descriptions of Clean Coal Technologies

## Washing Coal

Washing coal removes one form of sulfur (pyrite). Washing the coal also removes many of the minerals that would become ash. Typically, the coal is washed very near the coal mine before being shipped. The power plant then receives washed coal, which is free from any of the sulfur found in this first form.

## Scrubbers, including Flue Gas Desulfurization

The term "scrubber" is somewhat misleading. We are not actually scrubbing the coal. Rather, we are grabbing the SOx gas molecules from the general mix of combustion by-products before the SOx molecules reach the air.

Specifically, a scrubber is defined as any device, whether chemical or physical, which is used to capture sulfur by-products. There are many versions of scrubbers. The most commonly used form of scrubber is Flue Gas Desulfurization (FGD).

Flue Gas Desulfurization (FGD) involves spraying a limestone slurry into the gas stream. The slurry attaches to the sulfur, and the newly formed compound is removed for disposal. Flue Gas Desulfurization can remove 90% – 97% of SOx from the flue gases. In addition to reducing the amount of SOx, the slurry-SOx mixture can be transformed into gypsum (calcium sulfate) for use in the construction industry. Scrubbing by FGD also captures many other molecules, such as remaining fly ash, nitrogen compounds, and mercury compounds.

There are two types of FGD: dry FGD and wet FGD. Both are essentially the same. The only difference is the amount of water. The wet FGD uses more water, resulting in a soupy mixture. The dry FGD uses less water, and is usually the form of a spray.

## Low [temperature] NOx Burners

NOx molecules tend to form at higher temperatures (at approximately 2,500 degrees Fahrenheit). However, at lower temperatures, such as 1,400 to 1,700 degrees Fahrenheit, NOx is less likely to form. Low temperature NOx burners operate by burning the coal at a lower temperature, resulting in the formation of fewer NOx molecules.

Note that these burners are called simply Low NOx burners, but you should know that a lower burning temperature is what makes less NOx. Low NOx burners can reduce emissions of nitrogen oxides by up to 70%, depending on the type of coal used and the configuration of the furnace.

## Selective Catalytic Reduction (SCR)

The Selective Catalytic Reduction (SCR) and the Selective Non-Catalytic Reduction (SNCR) are used to reduce the amount of $NO_x$. In these methods Ammonia ($NH_3$) is then sent into the mix of by-products in order to react with the $NO_x$, creating harmless $N_2$ and $H_2O$.

The "catalyst" in the Selective Catalytic Reduction is usually a vanadium-titanium metal oxide ($V_2O_5$-$TiO_2$). The ammonia plus catalyst converts $NO_x$ into water ($H_2O$) and pure nitrogen ($N_2$). Water and pure nitrogen are both harmless, and both can be released into the air. Selective Catalytic Reduction (SCR) can remove 90% of the $NO_x$ molecules.

## Electrostatic Precipitator (to reduce ash)

The electrostatic precipitator captures the fly ash before the ash reaches the air. The electrostatic precipitator also removes mercury compounds. The electrostatic precipitator works by giving an electrical charge to the ash particles. Then a filter has an opposite charge, which draws the particles to the filter and keeps them there. The collection of ash particles can then be removed. The electrostatic precipitator has been around since 1950, and it can remove 99.7% of the fly ash.

## Circulating Fluidized Bed Combustion

In a fluidized bed, the coal pieces are held floating in the furnace, usually supported by a stream of hot air. The stream of hot air is what allows the coal pieces to float, and hence creates a "fluidized bed". Crushed limestone or dolomite is added to the mix. The limestone is a sulfur-absorbing chemical (same as the use of limestone in scrubbers). The new material is then collected along with the other ash. Circulating Fluidized Bed Combustion can capture 95% of the sulfur in the coal.

## Planting trees and other vegetation nearby

Trees, small plants, and agriculture naturally absorb the $CO_2$ in the air. If the $CO_2$ is captured, there will be no overall change in the amount of $CO_2$ in the atmosphere. In addition, this will provide oxygen, food, and habitat for many living things.

### Gasification

In Gasification, coal molecules are broken apart, then recombined to our liking. This process creates molecules which 1) are in the gas phase, and 2) can be separated easily. The pollutant gasses are separated and removed, while the desired fuel gasses are sent to be burned. The next section will discuss Gasification in detail.

# Gasification Process in Detail

### Introduction

Gasification will be a large part of our future regarding generating electricity from coal. Gasification in coal power plants is more environmentally friendly, reducing up to 99% of all pollutants. Gasification can also be more efficient, raising coal power plant efficiency from 40% to 70%. The benefits of gasification include: almost all Sulfur and Nitrogen are collected, most of the $CO_2$ is collected, and we create more electricity per mass of fuel.

The primary use for gasification is for coal. However, other raw materials can be used. Theoretically, anything that is carbon based can be a candidate for gasification. Some of the sources which will benefit from gasification include: bituminous coal, coal with high sulfur content, and biomass.

The following describes the gasification process in detail (with technical notes as necessary).

### Gasification Process in Detail

1. The molecular structure of coal is broken, so that the individual elements are free. The molecular bonds are broken by using high temperatures, and sometimes also with high pressures.

2. This process is done in an environment without much Oxygen. Having little or no oxygen is important so that the elements will combine in the way that we want.

3. The elements then recombine, but in ways more to our liking.

   Gas Phase – in large amounts
   a. CO   (Carbon Monoxide)
   b. $H_2$   (Elemental Hydrogen)
   c. $NH_3$  (Ammonia)
   d. $H_2S$, $H_2SO_4$   (Hydrogen Sulfide, Sulfuric Acid)

   Gas Phase – in small amounts
   e. $CO_2$  (Carbon dioxide)
   f. $CH_4$   (Methane)

   Solids
   g. Slag:  (ash which has been fused by the high temperatures)

5. The solids are taken away. These solids (the "slag") can be used in the construction industry. Therefore, there is no disposal problem.

6. The ammonia is separated easily and removed. In addition, this ammonia can be sold to industry.

7. The sulfur is separated easily and removed. The sulfur is taken in the form of hydrogen sulfide, pure sulfur, or sulfuric acid. Almost 99% of sulfur can be removed from coal in this manner. In addition to removal of these molecules, many of these sulfur products can be processed and sold to industry.

8. This leaves CO and $H_2$, both in the gas phase. This gas is called "syngas", which means "synthetic gas." This is relatively pure: up to 99% of all impurities have been removed at this point.

9. The burning now occurs. CO becomes $CO_2$. Some $H_2$ becomes $H_2O$. Some $H_2$ remains unreacted.

10. Then the remaining gases flow to the turbines and create electricity. Also note that we can use the "dual-turbine" system, and thus get more electricity.

11. The $H_2$ can be collected, for use in hydrogen fuel projects.

12. The $CO_2$ can also be collected. The gas stream is just $CO_2$ and water vapor. This makes the collection of $CO_2$ very easy.

## Modernization of Coal Power Plants

A coal power plant that is built today is very clean. However, older coal plants may not be so clean. These power plants must be modernized in order to produce power as cleanly as the power plants built today.

Coal power was one of our earliest forms of electrical power. Because of this, we have many power plants that are using old technology. The technology in a coal power plant can be anywhere from 10 years to 100 years behind the technology available today. These power plants must be modernized.

We have the available technology to make even the oldest of coal power plants very, very clean. The only obstacle to clean coal power on a national scale is our commitment to modernizing the older power plants with the latest clean coal technologies.

## Summary

1. Burning is the combination of any element with oxygen.

2. The primary by-products from burning coal may include:
    a. gas phase: $CO_2$, $H_2O$, $SO_x$, and $NO_x$.
    b. solid phase: $SiO_2$, $Al_2O_3$, $Fe_2O_3$, $CaO$, $MgO$, $Na_2O$, $K_2O$, and $TiO$.

3. Ash is any product from burning which is created in a solid form. The primary molecules found in ash include $SiO_2$, $Al_2O_3$, $Fe_2O_3$, and $CaO$.

4. Bottom ash and slag can be cleared out of the furnace easily. Fly ash can be separated and collected easily by an electrostatic precipitator. Disposing of ash is not a problem because ash can be used in other industries.

5. Most of the ash falls to the bottom of the furnace and is easily collected. Up to 99.7% of the fly ash is collected by using electrostatic precipitators. The remainder of the fly ash is collected by scrubbers. All the ash can be used for other purposes.

6. The major problem with $SO_x$ is acid rain. Sulfur is found with coal in two forms. One form is in pyrite, which is found with coal but not chemically attached. Sulfur is also bound chemically to the coal.

7. Sulfur can be removed by:
   a. Washing
   b. Scrubbers (Flue Gas Desulfurization)
   c. Gasification
   d. Circulating Fluidized Bed Combustion

8. Flue Gas Desulfurization removes 90%–97% of the $SO_2$.

9. The major problem with nitrogen is the by-product $NO_x$, which can become acid rain. As with the second form of sulfur, the nitrogen is chemically bound to the coal, and must be burned with the coal in order to remove. Nitrogen can be removed by:
   a. Low temperature NOx Burners (low temperature, thus low NOx)
   b. Selective Catalytic Reduction
   c. Gasification

10. Selective Catalytic Reduction (SCR) removes 90% of the $NO_2$ molecules.

11. The amount of mercury in coal is very small, only trace amounts. In many coal beds, mercury does not exist at all.

12. Clean Coal Technology (CCT) is any technology which reduces the amount of pollutants produced in the production of power from the burning of coal. The primary types of Clean Coal Technology include:

    a. Washing Coal

    b. Scrubbers, including Flue Gas Desulfurization (FGD)

    c. Low [temperature] NOx Burners (low temp., thus low NOx)

    d. Selective Catalytic Reduction (SCR)

    e. Electrostatic Precipitator

    f. Circulating Fluidized Bed Combustion (CFBC)

    g. Planting trees and other vegetation

    h. Gasification

13. Washing coal removes pyritic sulfur, as well as other minerals that would become ash.

14. A scrubber is any device which is used to reduce the amount of Sulfur by-products. Scrubbers also remove other by-products, such as fly ash and mercury. FGD is the most common scrubber. In FGD, a limestone slurry is sprayed into the gas stream. The slurry attaches to the sulfur, and the newly formed compound is removed for disposal.

15. Low NOx Burners operate at a lower temperature (1,700 degrees Fahrenheit versus 2,500 degrees Fahrenheit). At this lower temperature, burning of the coal can take place yet fewer NOx molecules are formed.

16. Selective Catalytic Reduction (SCR) is used to reduce the amount of NOx by-products. Ammonia is sent to the mixture, which converts NOx into water and pure Nitrogen.

17. The electrostatic precipitator separates the fly ash before reaching the atmosphere. The electrostatic precipitator works by giving an electrical charge to the ash particles, then a filter draws the particles and keeps them there.

18. In Circulating Fluidized Bed Combustion (CFBC) the coal pieces are held floating in the air. Limestone or dolomite is added to the mix which absorbs sulfur.

19. Planting trees reduces the amount of $CO_2$ which reaches the atmosphere, and therefore eliminates any rise in global temperature.

20. In gasification, coal molecules are broken apart then recombined to our liking. The pollutant gasses are separated and removed, while the desired fuel gasses are sent to be burned.

21. A coal power plant that is built today is clean. However, older coal must be modernized in order to produce power as cleanly as those power plants built today.

22. Gasification in coal power plants is more environmentally friendly, reducing up to 99% of all pollutants.

23. Gasification is more efficient than traditional coal power, raising efficiency from 40% to 70%.

24. In gasification: coal molecules are broken apart then recombined to our liking. This process creates molecules which are in the gas phase and can be separated easily. The pollutant gasses are separated and removed, while the desired fuel gasses are sent to be burned.

25. The main molecules we get from the process are carbon monoxide, elemental hydrogen, ammonia, hydrogen sulfide, sulfuric acid, carbon dioxide, methane, and slag.

26. The elements in coal (carbon, hydrogen, sulfur, and nitrogen), become the following in gasification:
   a. Carbon becomes: $CO$, $CO_2$, and $CH_4$
   b. Hydrogen becomes: $H_2$, $NH_3$, $H_2S$, $H_2SO_4$, and $CH_4$
   c. Sulfur becomes: $H_2S$ and $H_2SO_4$
   d. Nitrogen becomes: $NH_3$

27. The main benefits of gasification include:
   a. Almost no SOx produced.
   b. Almost no NOx produced.
   c. Almost all Sulfur and Nitrogen are collected.
   d. $CO_2$ collected, less greenhouse effect
   e. Coal with higher sulfur content can now be burned more cleanly.
   f. Two turbines and two electrical generators, resulting in more electricity per mass of fuel.

# 6.5
# $CO_2$ and the Environment

## Introduction

There has been debate regarding the effect of $CO_2$ on the environment, particularly regarding global temperature. In order to understand the issues we must look at the science, not the politics.

Note that I have personally spent many months researching this topic. I have gathered an enormous amount of data from a variety of sources. I researched any area of science that was related to carbon dioxide. I also researched any topic that was related to the temperature of the earth. What you read here is only a small fraction of the information I have acquired.

I also want to remind the reader that I have no agenda. The statements and conclusions I make here are based on extensive research and data analysis, not personal opinions. I wanted to be absolutely accurate when discussing this highly controversial topic. I stand behind my statements.

List of topics for this chapter
1. Global Heating Effect
2. Infrared Radiation and Greenhouse Gasses
3. Percentages of Greenhouse Gasses
4. Body Heat, $CO_2$ and Temperature
5. $CO_2$ and Plants
6. Volcanoes, Fires, and other Natural $CO_2$ Production
7. Earth History: Are We Just Going Through A Cycle?
8. Recent Temperature Trends

## The Global Heating Effect

The Global Heating Effect is a sophisticated process. The entire situation includes the sun, infrared waves, animals, rocks on the earth, and carbon dioxide in the atmosphere. We will discuss a few concepts here, however the subject is very complex, and much of the science is outside the scope of this book.

The primary source of temperature of the earth is the sun. Bursts of energy from the sun reach our planet, these bursts of energy hit molecules everywhere on the earth, and excite these various molecules. This energy is usually put into kinetic energy (molecules in motion). An increase in kinetic energy is in fact an increase in temperature. Therefore, because all these molecules on our planet are vibrating faster, the general temperature of our planet is higher.

The sun is the source of energy, yet the processes here on Earth regulate the temperature. The temperature of the planet at any given time or location depends on how the energy is transferred. Molecules transfer their temperature in various ways. The most common methods include bumping into other molecules, moving to other locations, or emitting electromagnetic energy. Note that these processes are done by plants, rocks, animals, and humans. These processes also occur with air molecules (including both greenhouse and non-greenhouse gasses). Through these processes, energy is transferred from place to place, resulting in a change in temperature in each region of the earth.

## Infrared Radiation and Greenhouse Gasses

The wavelengths known as infrared radiation are of particular concern to life on earth. This is because carbon molecules absorb infrared radiation, transforming electromagnetic energy into kinetic energy (and hence an increase in temperature of those molecules). All species of life on earth are carbon based, and therefore all life forms feel this energy as an increase in the internal temperature of their bodies.

Furthermore, many carbon molecules (including animals and rocks) emit infrared radiation (that is, many rocks and animals emit infrared as well as absorbing it). For all these reasons, infrared radiation is especially important to life on earth.

This leads us to the point of "greenhouse gasses." Greenhouse gasses are those gasses which act like a blanket with respect to infrared radiation. Without these greenhouse gasses, the infrared radiation would escape into space.

More specifically, the greenhouse gasses act like a sponge: absorbing some infrared radiation, vibrating, then re-radiating infrared radiation. (The reradiated infrared is emitted at a slightly lower energy). This energy is emitted in all directions. This includes toward space as well as toward earth.

Note that we do need some greenhouse gasses in our atmosphere. Scientists have estimated that without any greenhouse gasses the temperature of the planet would be approximately −18 degrees Fahrenheit. Without any greenhouse gasses, life on this planet would not be able to survive. Thus, carbon dioxide is necessary for life on earth.

The question that remains is what is the proper range of greenhouse gasses for life to thrive?

## Percentages of Greenhouse Gasses

The air around us mostly Nitrogen and Oxygen, yet neither of these molecules have anything to do with infrared. Greenhouse gasses are the only gasses which keep the infrared radiation from escaping into space. Yet greenhouse gasses are only 1% of all the gasses in the atmosphere. Is only 1% a relevant amount? The answer is: Yes.

The total re-radiation of infrared energy by all greenhouse gasses in the atmosphere is approximately 90%. Thus, if 1% of the gasses prevent 90% of the infrared energy from escaping into space, then a small change in the amount of those gasses, either too few or too many of those gas molecules, may have a significant effect on the temperature of the earth (and on the ability of life to survive).

Carbon dioxide is responsible for re-radiating about 84% of the infrared wavelengths in the atmosphere. There are a few other gases which have the same re-radiating effect that $CO_2$ does, including methane ($CH_4$) and nitrous oxide ($N_2O$). However, these exist in much smaller quantities than carbon dioxide. The amount of methane in the atmosphere is about 1/10 the amount of carbon dioxide, and the amount of nitrous oxide in the atmosphere is only 1/100 the amount of carbon dioxide. Therefore, the effects of the other greenhouse gasses are actually very minimal when compared to carbon dioxide.

## Body Heat, $CO_2$ and Temperature

The air around us mostly Nitrogen and Oxygen, yet neither of these molecules are affected by infrared. Therefore, other mechanisms for increase in air temperature must exist, and one of those mechanisms is body heat.

Many plants and animals emit body heat. You may be familiar with this phenomenon when you are in a crowded room and feel the temperature increase. How does this work? Molecules of plants and animals are constantly vibrating. When nearby air molecules hit the skin, these molecules absorb some of that energy.

Thus, the air molecules are now vibrating faster (and therefore these air molecules are existing at a higher temperature). Additionally, these higher temperature molecules bump into other molecules, or move to other locations, which transfers kinetic energy, and thus the temperature increases in other places. Compound this phenomenon by thousands of people or plants in an area, and the overall temperature of the surrounding air increases.

There are many ways in which a the internal temperature of a person or animal can increase. This energy will be transferred to surrounding air molecules when those molecules hit the skin.

Note that infrared radiation can be a primary cause of this temperature increase (in humans and in the surrounding air). The sequence would be as follows: The infrared radiation is absorbed by humans, and this energy sets molecules vibrating. Then air molecules hit the skin of the humans and therefore absorb that energy. Thus, the air molecules then vibrate faster, and exist at a higher temperature.

Humans and other animals can also emit various frequencies of electromagnetic energy, including infrared. Any of these emitted frequencies may be absorbed by nearby humans, animals, or objects nearby. When the human, animal, or object absorbs that energy, the molecules may vibrate faster, and hence the entity exists at a higher temperature.

## CO$_2$ and Plant Life

Carbon dioxide is good for all plant life. There are several studies which have shown that an increase in $CO_2$ actually helps trees grow. Studies of $CO_2$ and plant life have been done on a variety of trees, on a variety of small plants, on a variety of agriculture, and have been done in different locations around the world. All of these studies show that more $CO_2$ is good for plant life. All plants grow stronger, healthier, and sometimes faster, when given additional $CO_2$.

Note also that Carbon Dioxide is not a pollutant. In fact, Carbon Dioxide is like vitamin: it is a natural supplement which helps all vegetation grow stronger. While the public talks about sequestering $CO_2$, all plant life is hungry for more.

There is also evidence to support the theory that producing carbon dioxide is not as important a factor as the lack of trees. It is possible that we can indeed produce as much carbon dioxide as we want, as long as we have enough trees and other vegetation nearby to capture all the carbon dioxide that we produce.

## Volcanoes, Fires, and other Natural $CO_2$ Production

The production of $CO_2$ has existed long before coal power plants or automobiles. Natural events such as volcanoes, forest fires, and the breathing of animals have created $CO_2$ for millions of years. We must take these all these natural processes into account when discussing $CO_2$ and the environment.

A single volcanic eruption can create an enormous amount of $CO_2$, easily two to three times as much $CO_2$ in a month as a coal plant would produce in a year. A large forest fire, such as those which occur frequently in the Western United States, produces a great amount of $CO_2$. We must also not forget animals, including man, which exhale $CO_2$ as part of the respiratory process. Each of these natural processes contribute significant amounts of $CO_2$.

## Earth History: Are We Just Going Through A Cycle?

Before we make scientific judgments on the effects of our modern technology on the temperature, we should study the temperature of the earth over the past millions of years. Throughout the history of the earth, the global temperature has varied in large swings from high to low, and back to high. These temperature swings have occurred repeatedly.

The overall temperature generally ranges from 5 degrees Celsius to 17 degrees Celsius. Currently, the overall temperature of the earth is approximately 15 degrees Celsius (59 degrees Fahrenheit). Note that this current high is lower than the highs reached over the history of the planet.

The situation we experience today could simply be part of a natural earth cycle. The earth seems to go through a maximum temperature approximately every 100,000 years. For example, if we look at the data for average global temperatures back to 150,000 years ago, then we can see a trend that matches closely with today. (See graph below). Note that similar trends existed in prior years, such as the range of 250,000 years ago to 150,000 years ago.

The graph below shows the high and low global temperatures over the past 150,000 years. You will notice that the global warming trend started 20,000 years ago is almost identical to the warming trend started 150,000 years ago. The rate of increase in both periods is the same: 10 degree increase over 20,000 year period.

You will also notice that 130,000 years ago the temperature reached a high of 16 degrees Celsius. Our current temperature is 15 degrees. Clearly the earth has had greater temperatures prior to the existence of man, and we may see that high again before the warming trend stops.

You will also notice that 130,000 years ago the temperature reached a peak, and then decreased. Therefore, it is very likely that in our time the temperature will eventually reach a peak, and then begin to decrease. Thus, the global temperature is cyclic: the earth has seen increases in temperature before, and a cooling trend is likely to come soon.

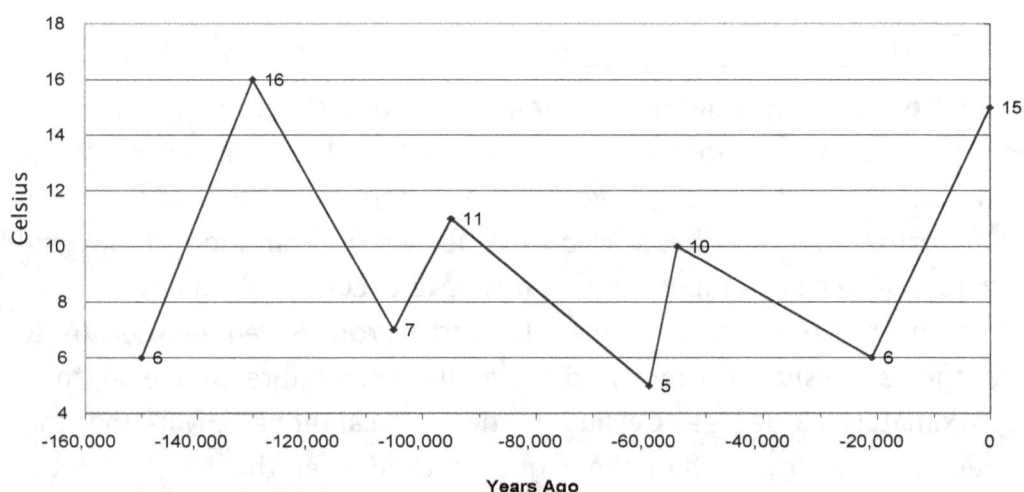

**Global Temperatures Past 150,000 Years**

# Recent Temperature Trends

## Introduction

The next question to be answered is: What are the recent temperature trends? The best way to answer this question is to look at the temperature data at a specific location (such as the state of Pennsylvania), for a specific month (such as June), over a period of years. Looking at the data for the same location, during the same month, year by year, we can then get an accurate understanding of temperature changes at that region of the earth.

It is important to think locally, not just globally, when we think of temperatures. This perspective is important for a variety of reasons. Foremost is that we must consider the effect of temperature on life in the area. The concern of warming is how it will affect life, including agriculture, animals, and people. Therefore, we must think locally, looking at local temperature trends, to predict how the temperature might alter the ecosystems of that area.

Second, different regions have different atmospheric and geographic conditions, and therefore temperature trends will not necessarily be equal in all areas. In other words, the weather in Chicago is often very different than the weather in Houston. Therefore, if we are to predict the effect of future temperatures on the ecosystem of an area, then we must think regionally, not globally.

## Data Overview

In the data below I have collected temperature data for four specific states: Iowa, Pennsylvania, California, and Texas. Each state has been chosen to represent a different area of the United States. Each data point is the average temperature of that state during the month of June. The temperature data spans over 100 years, from 1900 to 2008.

All temperature data comes from the National Climatic Data Center, which is the official record keeper of temperature in the United States. Similar data can be collected for other states and for other months.

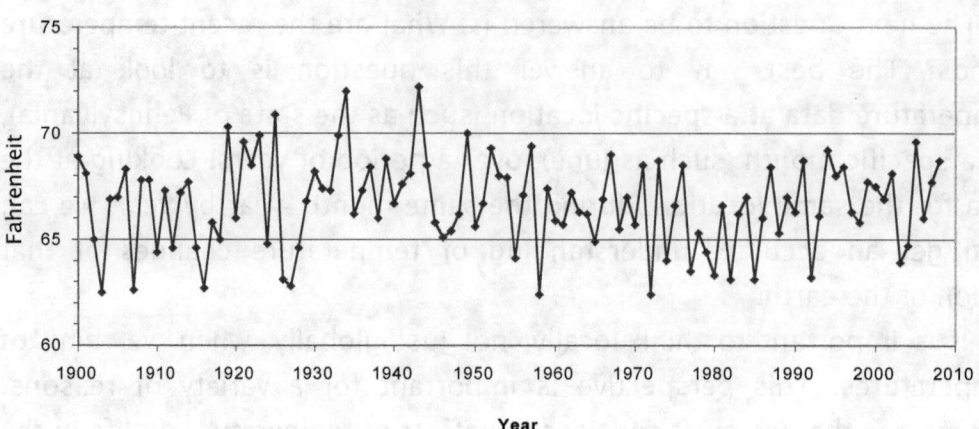

Pennsylvania Temperature Data 1900-2000
June of Each Year

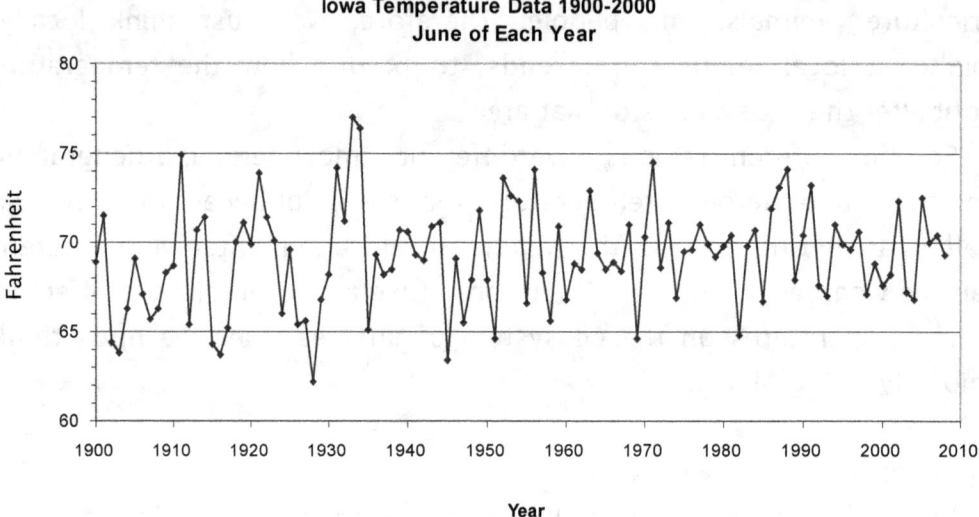

Iowa Temperature Data 1900-2000
June of Each Year

## Observations from the Data

From this data we can notice several things:

1. The temperature at a particular location on a particular date tends to fluctuate from year to year. The temperature is not constant.

2. The temperature can fluctuate as much as 10 degrees within a decade. In some cases the temperature can change as much as 10 degrees within a year or two.

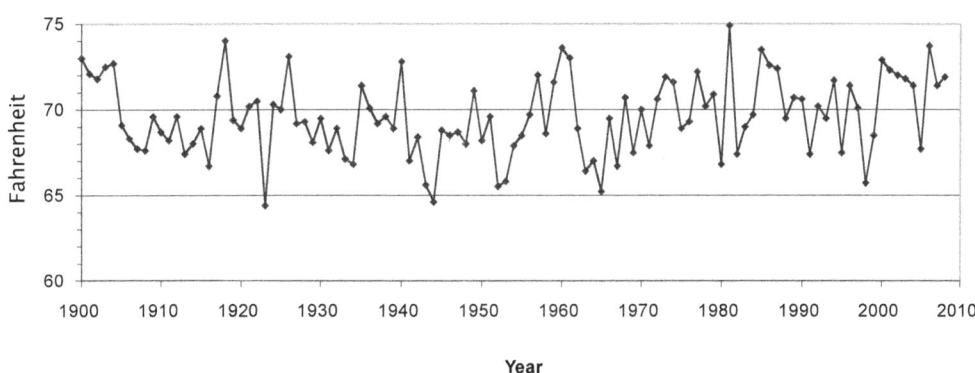

California Temperature Data 1900-2000
June of Each Year

3. There are just as many trends of decreasing temperature as increasing temperature. In other words, the temperature is cyclic. The temperature does not increase forever.

4. Temperatures in the period of 1900–1950 have been as high as (or higher than) the temperatures in the period of 1950–2008.

5. The most dramatic increase in temperature occurred in the Midwest, in the range of years 1928–1934. The temperature increased approximately thirteen degrees in eight years, or 1.5 degrees per year.

Notice however that the temperature *decreased* dramatically the following year (in 1935). In some locations the temperature dropped more than ten degrees in a single year.

6. Other similar dramatic increases in temperature have occurred throughout the century, yet each dramatic increase was followed by a significant decrease.

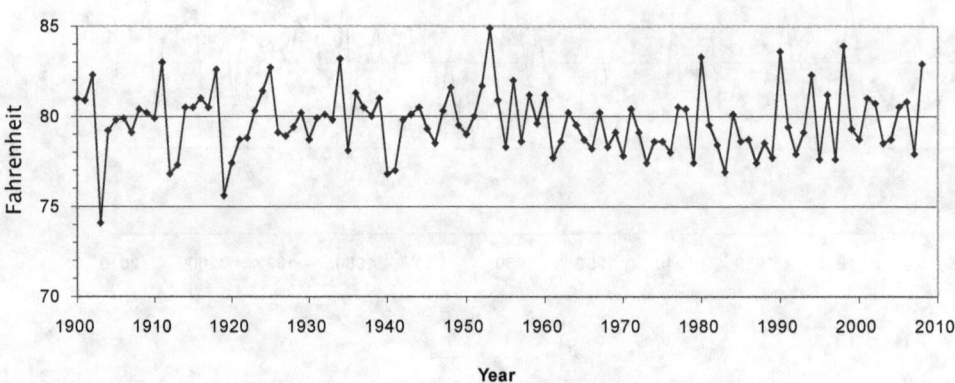

**Texas Temperature Data 1900-2000**
**June of Each Year**

7. There is no general warming trend in the United States. If there were a trend of irreversible increasing temperature then the data would show it. However, no such trend exists. The temperature goes down just as much as it goes up.

There may be warming trends in other parts of the world; I cannot speak to all areas of the earth. However, in the United States whenever the temperature has gone up, the temperature has always gone down again within a few years.

8. Regarding electrical power, there is no evidence that power plants have increased the temperature of the United States. Despite the number of coal power plants and the number of years they have been operating, the temperatures have not changed significantly.

The temperatures in the United States have increased and decreased cyclically within 5–10 year periods, and continue to do so. Therefore, at this time there seems to be no connection of coal power plants with the temperatures in the United States.

# Summary

1. The primary source of temperature of the earth is the sun. The sun's energy excites molecules all over the planet, resulting in an overall increase in temperature.

2. Infrared radiation is of particular concern to life on earth because all living creatures feel infrared radiation as an increase in temperature.

3. Greenhouse gasses are those gasses which act like a blanket with respect to infrared radiation. Without these greenhouse gasses, the infrared would escape into space.

4. We do need some greenhouse gasses in our atmosphere. Without any greenhouse gasses the earth would be too cold for us to survive.

5. Greenhouse gasses are only 1% of all the gasses in the atmosphere.

6. Carbon dioxide is responsible for re-radiating about 84% of the infrared wavelengths being sent from the earth.

7. There are a few other gases which have the same re-radiating effect that $CO_2$ does, including methane ($CH_4$) and nitrous oxide ($N_2O$). However, these exist in much smaller quantities.

8. Plants thrive on $CO_2$. Studies of $CO_2$ and plant life have been done on a variety of plant life in different locations around the world. All of these studies show that an increase in $CO_2$ results in stronger, healthier plants.

9. Natural events such as volcanoes, forest fires, and the breathing of animals have created significant amounts of $CO_2$ over millions of years.

10. Throughout the history of the earth, the global temperature has ranged from 5 degrees Celsius to 17 degrees Celsius. Currently, the overall temperature of the earth is approximately 15 degrees Celsius (59 degrees Fahrenheit).

11. Temperature data of the United States in the last hundred years shows a cycle of increasing and decreasing temperatures every few years.

# 6.6
# Mercury and Coal Power

## Introduction

Mercury from coal power has only recently become an issue. The main reason for the attention is that many of the other problems of coal power have been eliminated.

The amount of mercury emissions must be put into perspective. Coal does not contain much mercury. In most coal beds, the amount of mercury is listed simply as "trace." Furthermore, even the EPA admits that the amount of mercury emitted from coal power in the United States is only 1% of the mercury entering the environment worldwide.

The primary health concern from mercury is not any of the mercury emissions, but rather a secondary form of mercury called methyl mercury. Methyl mercury has the most serious health risks of all the forms of mercury.

Although mercury and its compounds have been studied for centuries, the study of mercury as it relates to coal power has only recently begun. Furthermore, the data is often scattered among various research organizations. Therefore, it is important that we collect and examine all the known science of mercury as it relates to coal power.

List of topics for this chapter
1. Amount of Mercury in Perspective
2. Mercury as Found in Coal
3. Mercury When Coal is Burned
4. Forms of Mercury in the Flue
5. Mercury After Being Emitted
6. Methyl Mercury – Basic Science
7. Minamata, Methyl Mercury, and Coal Power
8. Clean Coal Technology and Mercury: Introduction
9. Clean Coal Technology and Mercury: Existing Technology
10. Clean Coal Technology and Mercury: New Technology

# Amount of Mercury in Perspective

Introduction

The amount of mercury emitted by a coal power plant is actually very small. For example, an EPA report from 1978 concluded that "The study has found no evidence of a health or environmental problem as a result of emissions of mercury from power plants." The report further stated: "This conclusion has led to the recommendation that a specific control program to address mercury emissions from power plants is not necessary at this time." (See EIMS Report #50614: *An Assessment of Mercury Emissions from Fossil Fueled Power Plants"* by Goldgraben et al, EPA, 1978.)

In addition to the report cited above, there are other reports which have similar data. These reports show that the amount of mercury emitted by coal power is very small. The following information comes mostly from reports by the Environmental Protection Agency (EPA), the Center for Air Toxic Metals (CATM), the Energy and Environmental Research Center (EERC), and the Department of Energy's Information Administration (EIA). Note that these authorities are not beholden to coal power. The titles of many of these reports are listed in the bibliography.

Facts Related to Mercury and Coal Power

• In some coal beds, there is no mercury at all. Where coal beds have the largest concentrations of mercury, the amount of mercury is simply listed as "trace."

• The overall percentage of mercury in U.S. coal is only .0000075% (weight percent.) This means that 1 ton of coal, on average, contains only .0024 ounces (.068 grams) of mercury.

• The amount of mercury in the flue (after combustion of the coal) is only 5–10 grams of mercury per $m^3$ of space. (This is equivalent to only .005 – .01 grams of mercury per Liter of space).

• The amount of mercury released by coal power plants in the United States is only 1% of the mercury released into the environment worldwide.

•Volcanoes, oceans, and certain plants will naturally emit mercury into the air. Although these processes are not fully understood, scientists have estimated that natural processes emit at least 33% of all mercury emissions worldwide. (Remember that coal power in the U.S. emits only 1% of mercury worldwide.)

•The amount of mercury emitted by the industrial waste processes (municipal waste combustors and medical waste incinerators) emit as much or more mercury than coal power. Mercury emissions from various industrial processes, nationally, for the year 2000 are:
   •Municipal Waste Combustors emitted 50 tons of mercury
   •Medical Waste Incinerators emitted 47.5 tons of mercury
   •Coal Power Plants emitted 48 tons of mercury

Summary of Amount of Mercury in Perspective
   When we objectively look at the data, it is clear that the amount of mercury emitted by coal power plants is a relatively small amount.
   However, some forms of mercury can result in permanent neurological damage, if the dose is high enough. Therefore, in the rest of this chapter we will examine the chemistry of mercury as it relates to coal power.


# Mercury as Found in Coal

Mercury naturally occurs in many rocks and soils. The mercury gets to its current location due to a number of geological and biological processes. Only a few of these processes are understood at this time. Worldwide, mercury might exist naturally in many forms. However, in coal beds the mercury is found mostly in three forms. The forms of mercury found in a typical coal bed, in the relative order of amounts, are:
   1. Mercury Sulfide, $HgS$ (also called Cinnabar)
   2. Elemental Mercury, $Hg$
   3. Mercuric Chloride, $HgCl_2$

# Mercury When Coal is Burned

## Introduction

What mercury does in the furnace depends on the type of mercury. For our purposes, we can consider that there are two types of mercury entering the furnace: 1) elemental mercury, and 2) inorganic mercury.

## Elemental Mercury, Hg

Elemental mercury is the mercury used in thermometers. We know this form of mercury as a silver liquid. Scientists often write elemental mercury as $Hg^o$ in order to distinguish it from the common mercury ion, $Hg^{2+}$.

When $Hg^o$ is put into a furnace, this liquid mercury will become gas phase mercury: $Hg^o_{(gas)}$. Elemental mercury can also become mercury ion, $Hg^{2+}$. The processes are not completely understood. However, the result is important because $Hg^{2+}$ is easier to capture than $Hg^o$.

## Inorganic mercury / mercury salts

Inorganic mercury molecules include compounds such as $HgS$ and $HgCl_2$. Inorganic mercury molecules are referred to by several general terms, including: inorganic mercury, mercury compounds, and mercury salts. When inorganic mercury molecules are heated in the furnace, the atoms separate. The mercury from inorganic compounds exists in the flue as $Hg^{2+}$.

# Forms of Mercury in the Flue

## Introduction

We need to know the specific mercury molecules that exist, and where they form, so that we can capture mercury more effectively. Different types of mercury have different properties, and therefore will require different capturing techniques.

The temperature is a critical factor in which types of mercury will exist. Some mercury compounds will form only when the temperature is cool enough. Note that the temperature will change at different points along the flue, therefore different types of mercury will form at different locations.

The by-products formed by mercury can be either a gas or a solid.

The types of mercury in the flue are most likely to be: Elemental Mercury, Hg; Mercuric Sulfide, HgS; Mercuric Chloride, $HgCl_2$; or Mercury Ion, $Hg^{2+}$.

Elemental Mercury, Hg, is very likely to exist. Mercury can enter the furnace as elemental mercury, and therefore elemental mercury can be emitted at the end. Mercuric Sulfide, HgS, is very likely to exist. Mercury can enter the furnace as HgS, and therefore even when the bonds are broken, the HgS can easily form again. Note that mercuric sulfide is the most common source of mercury for commercial purposes.

Mercury Chlorides, $HgCl_2$ and $Hg_2Cl_2$ are very likely to exist. If there is any chlorine in the flue from any source, the mercury ion will easily bond with the chlorine. Note that adding Chlorine, such as by adding HCl, is a proposed technique for capturing mercury.

Note that Mercuric Oxide, HgO, is not likely to exist. The temperature is too high for HgO molecules to form. Mercuric Sulfate, $HgSO_4$ is not likely to form. The temperatures in the flue are usually too high for $HgSO_4$ to form. Mercuric Nitrate, $Hg(NO_3)_2$, is not likely to form. As above, the temperatures near the furnace are too high for $Hg(NO_3)_2$ to form.

Dimethyl Mercury, $(CH_3)_2Hg$, is not likely to exist. Carbon and Hydrogen prefer to react with Oxygen, forming $CO_2$ and $H_2O$. Therefore, the Carbon and Hydrogen are not likely to be available for $(CH_3)_2Hg$ to form. Furthermore, the primary means of creating dimethyl mercury is through an anaerobic bacteria. (This process is explained in greater detail in a later section).

# Mercury After Being Emitted

## Introduction

When discussing the relative health hazards of each type of mercury, we first divide the forms of mercury into three general categories: elemental, inorganic, and organic.

1. Elemental mercury is toxic. It can be inhaled or absorbed through the skin. However, elemental mercury as emitted from coal power plants is a relatively low health hazard due to the low amount of elemental mercury produced.

2. Inorganic mercury salts are toxic. The toxicity varies depending on the specific type of molecule.

The hazards of inorganic mercury also depend on the concentration in an area: if the local weather manages to deposit a sizable quantity of inorganic mercury in one location, then the amount of inorganic mercury ingested by local animals may be significantly high. On the other hand, if local weather manages to spread the inorganic mercury molecules over many miles, then the amount ingested by any animal will be relatively low, and therefore be well within safe limits.

It is best to place instruments which measure the quantities of inorganic mercury throughout the region. (Local colleges, ranches, and weather stations are ideal locations).

3. Organic mercury is not created by coal power, however the mercury emitted from the coal power plant can be converted into organic mercury by bacteria. There are several forms of organic mercury, but the most common is dimethyl mercury.

# Methyl Mercury – Basic Science

## Introduction

The most important of all mercury compounds in relation to human health are a series of related compounds known as "methyl mercury." However, methyl mercury compounds are not produced in coal power plants.

There are several forms of methyl mercury. Forms of methyl mercury include: $(CH_3)_2Hg$, $(CH_3)HgCl$, and $(CH_3)_3HgHCl$. Of these forms, dimethyl mercury $(CH_3)_2Hg$ is the most common. Therefore, dimethyl mercury is the methyl mercury compound we will discuss most

Methyl mercury is often abbreviated as MeHg. However, this abbreviation is not technically accurate, and should not be used. Also note that the specific forms of methyl mercury, including dimethyl mercury, are often simply referred to as methyl mercury. However, it is best to identify the specific forms of methyl mercury compounds. Specific identification is useful because each type differs in formation process, differs in toxicity, and differs in methods of capture.

## Formation of Methyl Mercury

Methyl mercury compounds are created by anaerobic bacteria which feed off the mercury. These anaerobic bacteria will feed on elemental mercury or on any of the inorganic mercury salts. Different types of bacteria will produce different forms of methyl mercury. Also note that the amount of methyl mercury produced depends on many factors, including specific type of bacteria and pH level.

Most of the methyl mercury created in the last 100 years was created by industrial processes, and usually in the specific form of dimethyl mercury. Note that most of these industrial processes have been eliminated world-wide. Also note that coal power plants have never created any form of methyl mercury.

## Health Concerns of Methyl Mercury

Methyl Mercury primarily affects the nervous system. The result can include loss of motor control, brain damage, and loss of senses (blindness, numbness). Methyl mercury can also affect the respiratory system and gastrointestinal systems. If the concentration of methyl mercury is extremely high, then death will result.

One of the problems with methyl mercury is that is easily ingested. There are several reasons: 1) methyl mercury is soluble in fat, 2) methyl mercury can pass through cells easily, and 3) methyl mercury is primarily ingested by fish, which are eaten by many people.

The amount of dimethyl mercury in a fish depends on several factors. It is not possible to generalize. Factors include: temperature, pH level, weather patterns, types of microorganisms in the soil and water, types of fish in the ecosystem, specific fish eaten by people, rate of mercury particle deposition, and forms of the mercury molecules. The EPA, the FDA, and various State Agencies have been studying the methyl mercury level in various fish, for different regions of the country. Their data is available to the public through various websites.

The safe level of mercury consumption varies depending on whose report you read. Daily consumption values of mercury are listed between .1 microgram per kilogram body weight daily, and 1.6 microgram per kilogram body weight daily. For a 150 pound person, this range of mercury consumption would be between 6.8 micrograms and 109 micrograms daily.

# Minamata, Methyl Mercury, and Coal Power

The case which started the investigation of methyl mercury was in Minamata, Japan, in the 1950s. In Minamata, a significant number of the people had symptoms which indicated degeneration of the nervous system. The cause was determined to be dimethyl mercury.

The primary source of the dimethyl mercury was a chemical factory which dumped dimethyl mercury waste into the local waters. Fish consumed the dimethyl mercury, then the local Japanese ate the fish, and because the people of Minamata lived entirely off local fish, the entire community ingested dimethyl mercury. It is generally estimated that the Japanese company dumped between 30 and 150 tons of mercury compounds into Minamata Bay and nearby waters.

Note that the Minamata case differs from the situation of coal power in the United States:

1. In the Minamata case, dimethyl mercury was a direct by-product of the chemical process. In contrast, coal power plants do not create any forms of methyl mercury.

2. In Minamata, the chemical factory dumped dimethyl mercury waste directly into the local waters. In contrast, coal plants in the United States do not dump mercury waste anywhere.

# Clean Coal Technology and Mercury: Introduction

Any technology proposed to capture mercury must be designed specifically for each form of mercury. This is because the properties of each form are slightly different. When designing a technology to capture mercury we can divide the forms of mercury into three general categories:

1. Elemental Mercury, $Hg^0$. ($Hg^0$ exists in the vapor phase)
2. Inorganic Mercurial Salts. (These include $HgS$ and $HgCl_2$)
3. Ionic Mercury, $Hg^{2+}$. (These float around, unattached)

# Clean Coal Technology and Mercury: Existing Technology

## Introduction

The existing Clean Coal Technologies (as used to reduce other pollutants) will capture some of the mercury. Current clean coal technologies will remove approximately 60% of the mercury.

Remember that there are three main existing clean coal technologies: Flue Gas Desulfurization (used to remove Sulfur), Selective Catalytic Reduction (used to remove Nitrogen), and Electrostatic Precipitators (used to remove fly ash). In this section we will look at how each of these technologies can remove some of the mercury.

## FGD, Flue Gas Desulfurization

Flue Gas Desulfurization is effective in removing $HgS$, $HgCl_2$ and $Hg^{2+}$. However, Flue Gas Desulfurization will not remove elemental mercury.

One of the reasons that FGD works so well in removing mercury salts is that inorganic mercury molecules are soluble in water. Remember that FGD works by using a slurry of water and limestone. Molecules such as $HgS$ and $HgCl_2$ dissolve in the water. These dissolved molecules are then absorbed by the limestone. Similarly, the mercury ion $Hg^{2+}$ is very soluble in water. Therefore, Flue Gas Desulfurization is effective in removing the mercury ion $Hg^{2+}$.

Note that elemental mercury is not soluble in water. Therefore elemental mercury cannot be captured using FGD. However, if the elemental mercury can be ionized into $Hg^{2+}$, then that mercury can be absorbed during the Flue Gas Desulfurization process.

Flue Gas Desulfurization is very effective in removing Mercury Sulfide. This is because FGD is used to remove Sulfur, and Mercury Sulfide has sulfur, therefore Flue Gas Desulfurization is effective in removing Mercury Sulfide.

## Selective Catalytic Reduction (SCR)

The SCR process turns some of the elemental mercury into ionized mercury, $Hg^{2+}$. This can be useful because $Hg^{2+}$ can be captured by Flue Gas Desulfurization. Therefore, by sending the mercury through SCR first, more of the mercury can be absorbed during FGD. Remember that Selective Catalytic Reduction changes the form of mercury, but does not capture any.

## Electrostatic Precipitator

Remember that the electrostatic precipitator is designed to remove small ash particles. Inorganic mercury, such as HgS and $HgCl_2$, are also small particles. Therefore, the electrostatic precipitator can be very effective in removing inorganic mercury particles.

# New Clean Coal Technology for Mercury

## Introduction

In order to remove additional mercury from the flue, new technologies must be developed. The most promising of the new technologies is using a sorbent called activated carbon. Activated carbon is essentially a carbon particle which is used to capture unwanted molecules.

You might also hear this activated carbon particle referred to as "activated charcoal" or as a type of "sorbent." The process also goes by several terms including "activated carbon injection", "sorbent capture", and "carbon filtration."

The use of activated carbon has been around for many years. Activated carbon is most commonly used as a method for purifying water.

## Sorbents, Absorb, and Adsorb

In general, a "sorbent" will capture unwanted molecules. Note that the term "sorbent" can apply to many different substances. Note also that different sorbents are used for different applications. Regarding the capture of mercury in a coal power plant, the sorbent being developed is activated carbon.

There are two ways in which a sorbent may capture unwanted molecules: absorb or adsorb. The difference is whether the unwanted molecules are captured on the inside of the sorbent (like a sponge), or the unwanted molecules stick to the surface of the sorbent (like sticky tape).

A sorbent which *absorbs* will work like a sponge. In the absorption case, the sorbent has pores, and the unwanted molecules are captured inside these pores. A sorbent which *adsorbs* will work like sticky tape. In the adsorption process, the surface of the sorbent will hold onto unwanted molecules. When the unwanted molecule hits the surface of the sorbent, that unwanted molecule will stick. Note that activated carbon uses both absorption and adsorption to capture the unwanted molecules.

## Adsorption and Difficulties in Sticking

Adsorption can be a difficult technology. The whole premise is that the unwanted molecule must stick to the sorbent. But this can be tricky. The molecule may hit the activated carbon but bounce off without sticking. Furthermore, a molecule which does stick to the surface might come off again before the activated carbon is collected.

The effectiveness of adsorption depends on three main factors: surface area of sorbent, surface geometry of sorbent, and temperature.

## Activated Carbon is Not Ready for Practical Use

Although activated carbon injection (ACI) is promoted as the great answer to mercury capture, the research papers conclude that we are not ready to use activated carbon for capturing mercury at this time. There are still many practical issues to overcome before activated carbon injection will be practical in coal power plants. Some of these issues are described below.

Note that the first power plant facility with an activated carbon capture system has recently been built. Researchers at this plant will be improving the activated carbon capture system, and hope to have the system perfected within the next few years.

Making the Activated Carbon work Effectively

The principle of using activated carbon to collect unwanted molecules is relatively simple. However, the chemistry involved and the mechanical engineering issues make the actual implementation more complex. Some of the primary factors of using activated carbon effectively include:

1. Pore size
2. Cavern size
3. Surface area
4. Concentration of mercury in the flue
5. Diameter of the pipe
6. Total number of carbon particles
7. Effect of other molecules in the flue
8. Stage in the sequence of clean coal technology processes
9. Temperature

## Phytoremediation and Microorganisms

A new approach to reducing the risk of dimethyl mercury poisoning is a process called phytoremediation. In phytoremediation, a plant attracts and absorbs dimethyl mercury. These plants can then be collected, removing dimethyl mercury from the area. Some plants are being developed which not only absorb the dimethyl mercury, but convert it into a safer form such as $Hg°$. This process will take the dimethyl mercury out of the food chain altogether.

There are also microorganisms which feed on dimethyl mercury. These microorganisms convert dimethyl mercury into mercury sulfide, which is essentially not a health problem and is considered safe. These microorganisms provide a possible solution for the dimethyl mercury issue.

# Chapter Summary

1. Mercury emissions from coal power have only recently become an issue. The main reason for this is that most of the other problems of coal power have been eliminated.

2. Coal does not contain much mercury. The overall percentage of mercury in U.S. coal is only .0000075%. The mercury emitted from coal power in the United States is only 1% of the mercury entering the environment worldwide.

3. Mercury in coal beds is found mostly in three forms: Mercury Sulfide, (HgS), Elemental Mercury (Hg), and Mercuric Chloride ($HgCl_2$)

4. When mercury is burned the primary resulting forms are elemental mercury (Hg) vapor, and ions of $Hg^{2+}$

5. We need to know the specific mercury molecules formed because the techniques to capture mercury will depend on the specific molecules in the flue. However, we don't know for certain which molecules form. Research is being done at this time.

6. The by-products formed by mercury can be either a gas or a solid.

7. Temperature and the other chemicals in the flue affect which molecules of mercury will form.

8. The most likely forms of mercury in the flue are:
   a. Elemental Mercury, Hg
   b. Mercuric Sulfide, HgS
   c. Mercuric Chloride, $HgCl_2$
   d. Mercury Ion, $Hg^{2+}$

9. The real threat to health from mercury is not elemental mercury nor inorganic mercury but rather organic mercury. The form of greatest health risk is dimethyl mercury.

10. The most important of all mercury compounds in relation to human health is dimethyl mercury, $(CH_3)_2Hg$. Although not produced in coal power plants, dimethyl mercury is produced naturally from many forms of mercury, given the proper conditions.

11. Dimethyl mercury is created by anaerobic bacteria which feed off the mercury. These anaerobic bacteria will feed on elemental mercury or on any of the inorganic mercury salts.

12. There are other microorganisms which feed off of the dimethyl mercury, and convert dimethyl mercury into mercury sulfide. Mercury sulfide is essentially not a health problem, and is thus considered safe. These microorganisms provide a possible solution for the dimethyl mercury issue.

13. Methyl mercury primarily affects the nervous system. Methyl mercury can also affect the respiratory system and gastrointestinal systems.

14. If a technology is being designed to capture mercury, then that technology must be designed to capture a particular type of mercury. When designing technologies to capture mercury, the mercury is categorized into three types:
   a. Elemental mercury, Hg, as vapor
   b. Inorganic Mercurial Salts (such as HgS and $HgCl_2$)
   c. Ionic Mercury, $Hg^{2+}$

15. Flue Gas Desulfurization is effective in removing HgS, $HgCl_2$ and $Hg^{2+}$. However, FGD will not remove elemental mercury.

16. Selective Catalytic Reduction turns some of the elemental mercury into ionized mercury, $Hg^{2+}$. The $Hg^{2+}$ can then be captured by Flue Gas Desulfurization.

17. The electrostatic precipitator can be very effective in removing inorganic mercury particles.

18. The most promising of the new technologies for capturing mercury is using activated carbon injection (ACI). However, we are not ready to use activated carbon at this time. There are still many practical issues to overcome before activated carbon injection will be practical in coal power plants.

19. A new approach to reducing the risk of dimethyl mercury poisoning is a process called phytoremediation. In this process a specific type of plant attracts, absorbs, and in some cases detoxifies the dimethyl mercury. These plants can then be collected and removed from the food chain.

# Conclusion

Many Americans hold passionate views about electrical power, yet few Americans understand all the details behind their passion. Electricity should not be mysterious. The science, the technology, and the data of electrical power can be understood by anyone.

Above all else, we must remember that there are no perfect solutions, there are only choices. Any option can be beneficial, yet each option has its own technical issues to work with. It is up to you and to your community to make those educated decisions. I hope that this book will help guide you in your choices.

M.F.

# Ash Data

| Molecule | Weight % of molecule in the ash |
|---|---|
| 1. $SiO_2$ (Silicon Dioxide) | 30% –50% |
| 2. $Al_2O_3$ (Aluminum Oxide) | 10% –25% |
| 3. $Fe_2O_3$ (Iron oxide) | 5% –10% |
| 4. CaO (Calcium Oxide) | 2% –25% |
| 5. MgO (Magnesium Oxide) | 1% – 6% |
| 6. $Na_2O$ (Sodium Oxide) | .2%–1.5% |
| 7. $K_2O$ (Potassium Oxide) | 1%–3% |
| 8. TiO (Titanium Oxide) | 1% |

Note that the exact percentages depend on the region where the coal came from. The percentages above are the averages for coal ash from all regions in the United States.

# Bibliography

## Coal Power

1. Energy for Man: From Windmills to Nuclear Power, by Hans Thirring, 1958. Indiana University Press.

2. Energy Resources, by Andrew Simon, 1975. Pergamon Press, Inc.

3. Nontechnical Guide to Energy Resources, by Ben Ebenhack, 1995. PennWell Publishing Company

4. Electric Power Generation: A Nontechnical Guide, by Barnett and Bjornsgaard, 2000. PennWell Publishing Company

5. Energy: A Guidebook, by Janet Ramage, 1997. Oxford University Press.

6. Coal Mines, video, part of the "Modern Marvels" series. A&E Productions.

7. American Coal Foundation www.acf-coal.org

8. World Coal Institute www.wci-coal.com/web/bl_content.pho?menu_id=0.0

9. Illinois Clean Coal Institute (ICCI) www.icci.org/index.html

10. Kentucky Coal Education Web Site www.coaleducation.org

11. Coal Utilization Research Council www.coal.org/index.html

12. United Mine Workers Association (UMWA) www.umwa.org

13. Coal Mining Engineer's Webpage http://home.inter.net/takakuwa

14. Handbook for Dust Control in Mining, NIOSH, 2003.

15. Mine Safety and Health Administration (MSHA) www.msha.gov

16. NIOSH (National Institute for Occupational Safety and Health) www.cdc.gov/niosh

17. Office of Fossil Energy (in Dept of Energy) http://www.fossil.energy.gov

18. National Energy Technology Laboratory (NETL), Office of Coal and Environmental Systems (in Dept of Energy) www.netl.doe.gov/coalpower

19. Gasification Technologies Council www.gasification.org

20. Ash Library www.flyash.info

21. American Coal Ash Association (ACCA) www.acaa-usa.org

22. The Fly Ash Resource Center www.geocities.com/CapeCanaveral/Launchpad/2095/flyash.html

23. Chemistry in Context, Second Edition, by the American Chemical Society (various contributors), 1997. Publisher: American Chemical Society

24. Center for the Study of Carbon Dioxide and Global Change, http://www.co2science.org

25. Carbon Dioxide and Plant Growth, Center for Study of Carbon Dioxide and Global Change http://www.co2science.org/co2tables/plantgrowth.htm

26. "How Did Humans First Alter Global Climate, by Dr. William Ruddiman, *Scientific American*, March 2005.

27. ICWG Report: *"Ice Core Contributions to Global Change Research,"* by the Ice Core Working Group (ICWG), May 1998.

28. ICWG Report: *"U.S. Ice Core Sciences – Recommendations for the Future,"* by the Ice Core Working Group (ICWG), June 2003.

29. American Electric Power (AEP) www.aep.com/default.asp

30. Earth Vision www.earthvision.net/ColdFusion

31. Mercury Answers www.mercuryanswers.org

32. EPA Mercury pages www.epa.gov/mercury

33. EPA Report: *"Control of Mercury Emissions from Coal–Fired Electric Utility Boilers"*, EPA, Air Pollution Prevention Division, 2004.

34. EPA Report: *"An Assessment of Mercury Emissions from Fossil Fueled Power Plants,"* by Goldgraben, Krickenberger, Clifford, Zimmerman, and Martin, EPA, 1978.

35. Center for Air Toxic Metals (part of Energy & Environmental Research Center) http://www.undeerc.org/catm

36. CATM/EERC Report: *"Development of Mercury Control Technologies,"* by Miller, Holmes, Dunham, Olson, Almlie, and Pavlish, for the Center for Air Toxic Metals/EERC, 2002.

37. CATM/EERC Report: *"Fundamental Study of SCR Impact on Mercury Speciation,"* by Laudal, Dunham, Brickett, for the Center for Air Toxic Metals/EERC, 2002.

38. CATM/EERC Report *"Development of Sampling and Analytical Tools for Oxidized Mercury Species,"* by Olson, Thompson, and Sharma, for the Center for Air Toxic Metals/EERC, 2001.

39. CATM/EERC Report *"Fundamental Mechanisms of Mercury Species Formation in Coal Combustion Flue Gas,"* by Zygarlicke, Galbreath, and Toman, for the Center for Air Toxic Metals/EERC, 2001.

40. Toxicology: The Basic Science of Poisons, 4th edition, edited by Cassarett, Doull, Ambur, and Klassen, 1993. Publisher: McGraw–Hill.

41. Minamata Disease: The History and Measures, Ministry of the Environment, Japan. http://www.env.go.jp/en/topic/minamata2002/index.html

42. Case Studies: Minamata Disaster, Trade and Environment Database (TED), American University, http://www.american.edu/TED/MINAMATA.HTM

43. Epidemiology of Minamata Disease, by Takizawa and Seikikawa, University of Pittsburgh Lecture, www.pitt.edu/~super1/lecture/lec0361/index.htm

44. *"Characteristics of Hg–resistant bacteria isolated from Minamata Bay sediment."* by Nakamura, Fujisaki, and Tamashiro, 1986. Report accessed via National Library of Medicine

45. *"Mercury dispersion from Minamata Bay to the Yatsushiro Sea during 1975– 1980."* by Kudo and Miyahara, 1984. National Library of Medicine.

46. *"The Engineered Phytoremediation of ionic and methylmercury pollution"* by Richard Meagher, Dpt. of Genetics, University of Georgia, 2000.
47. *"Our Preferred Poison,"* by Karen Wright, Discover, March 2005.

48. *"U.S. Coal Supply and Demand: 2003 Review"* by Fred Freme, Energy Information Administration, April 2004.
49. Annual EIA Report: *"Annual Coal Report 2003, Executive Summary"*, by Fred Freme, Energy Information Administration, last updated January 2005.

## Department of Energy (DOE) Related Sites

1. Department of Energy (DOE)  www.energy.gov
2. Energy Information Administration  (EIA) www.eia.doe.gov
3. [Office of] Efficiency and Renewable Energy (EERE)  www.eere.energy.gov
4. Office of Fossil Energy (in Dept of Energy) www.fossil.energy.gov
5. Electric Transmission and Distribution Office  www.electricity.doe.gov
6. Science (Office of Science) www.sc.doe.gov
7. Nuclear Regulatory Commission (NRC)  www.nrc.gov
8. Civilian Radioactive Waste Management (OCRWM)  www.ocrwm.doe.gov
9. Yucca Mountain Project  www.ocrwm.doe.gov/ymp/about/index.shtml
10. International Nuclear Safety Program  http://insp.pnl.gov
11. International Nuclear Safety Center, Argonne Laboratory  www.insc.anl.gov
12. National Energy Technology Laboratory (NETL) www.netl.doe.gov
13. National Renewable Energy Laboratory (NREL) www.nrel.gov
14. Oak Ridge National Laboratory  www.ornl.gov
15. Los Alamos National Laboratory (LANL) www.lanl.gov/worldview
16. Pacific Northwest National Laboratory (PNL)  www.pnl.gov
17. Starlight, from  PNNL/DOE  http://starlight.pnl.gov

## Energy & Environmental Research Center (EERC)

1. Energy & Environmental Research Center (at Univ. of North Dakota) www.undeerc.org
2. Coal Ash Resource Center www.undeerc.org/carrc/index.html
3. Center for Air Toxic Metals (CATM)  www.undeerc.org/catm
4. Center for Biomass Utilization (CBU) www.undeerc.org/centersofexcellence/biomass/default.asp

## Government Sites – General

1. <u>US Department of Energy (DOE)</u>  www.energy.gov
2. <u>US Department of the Interior</u>  www.doi.gov
3. <u>US Bureau of Reclamation</u> www.usbr.gov
4. <u>US Department of Agriculture (USDA)</u>  www.usda.gov
5. <u>Environmental Protection Agency (EPA)</u>  www.epa.gov
6. <u>Food and Drug Administration (FDA)</u>  www.cfsan.fda.gov
7. <u>National Institute for Occupational Safety and Health (NIOSH)</u>
   www.cdc.gov/niosh
8. <u>Mine Safety and Health Administration (MSHA)</u> www.msha.gov
9. <u>Federal Energy Regulatory Commission (FERC)</u>  www.ferc.gov
10. <u>Nuclear Regulatory Commission (NRC)</u>  www.nrc.gov
11. <u>National Climatic Data Center (NCDC)</u> www.ncdc.noaa.gov

# Index

www.ingramcontent.com/pod-product-compliance
Lightning Source LLC
Chambersburg PA
CBHW081240180526
45171CB00005B/488